Semiosis and Catastrophes

European Semiotics: *Language, Cognition, and Culture*
Sémiotiques Européennes: *langage, cognition et culture*

Edited by / Série rédigée par
Per Aage Brandt (Cleveland), Wolfgang Wildgen (Bremen/Brême),
and/et Barend van Heusden (Groningen/Groningue)

Volume 10

PETER LANG
Bern · Berlin · Bruxelles · Frankfurt am Main · New York · Wien

Wolfgang Wildgen &
Per Aage Brandt (eds.)

Semiosis and Catastrophes

René Thom's Semiotic Heritage

PETER LANG
Bern · Berlin · Bruxelles · Frankfurt am Main · New York · Wien

Bibliographic information published by Die Deutsche Bibliothek
Die Deutsche Bibliothek lists this publication in the Deutsche Nationalbibliografie; detailed bibliographic data is available on the Internet at ‹http://dnb.ddb.de›.

A catalogue record for this book is available from *The British Library*, Great Britain.

Library of Congress Cataloging-in-Publication Data

Semiosis and catastrophes : René Thom's semiotic heritage / Wolfgang Wildgen & Per Aage Brandt (eds.).
p. cm. – (European semiotics: language, cognition, and culture sémiotiques Européennes: langage, cognition et culture; v. 10.)
Papers in English and French.
Most of the papers in this volume are results of this conference. Università degli Studi di Urbino Carlo Bo, Centro Internazionale di Semiotica e Linguistica, where the conference L'eredità semiotica di René Thom, Semiosi e catastrofi had been organized from the 18th to the 20th of July 2005.
ISBN 978-3-0343-0467-2
1. Thom, René, 1923–Congresses. 2. Semiotics–Congresses. 3. Modality (Linguistics)–Congresses. 4. Catastrophes (Mathematics)–Congresses. I. Wildgen, Wolfgang. II. Brandt, Per Aage.
P99.S36886 2010
401'.41–dc22

2010037755

ISSN 1423-5587
ISBN 978-3-0343-0467-2

Hochfeldstrasse 32, CH-3012 Bern
info@peterlang.com, www.peterlang.com, www.peterlang.net

Printed in Switzerland

Acknowledgments

Special thanks to the Università degli Studi di Urbino "Carlo Bo", Centro Internazionale di Semiotica e Linguistica, where the conference "L'eredità semiotica di René Thom, Semiosi e catastrofi" had been organized from the 18th to the 20th of July 2005 (by Per Aage Brandt and Wolfgang Wildgen). Most of the papers in this volume are results of this conference.

Cet ouvrage a bénéficié d'une aide du programme "Foundations of Physics" de l'ANR (Agence Nationale de la Recherche) de Paris.

Contents

Introduction

WOLFGANG WILDGEN

1. Intellectual biography of René Thom[1]

René Thom was born is Montbéliard (France) in 1923 (2nd of September). Until 1940 he lived there with his parents, his brother and several grand parents. In 1940 the brothers went to Switzerland to escape the war, but they were brought back to Lyon, where René Thom passed his "baccalauréat. In 1941 he entered the preparatory classes of the Lycée Saint-Louis in Paris and was inscribed in "Higher mathematics and special mathematics".[2] His favored professor was teaching ancient Greek; this remained a center of intellectual concern during his life. In 1943 he entered the elite college ENS (École Normale Supérieure). After some difficulties at the end of the war (he was forced to work for the German troops to fortify Belfort) he returned to the ENS in Paris and then went to Strasbourg to finish his "Agrégation des sciences mathématiques" (his teacher was Henri Cartan) in 1946.[3] Between 1947 and 1951 he was a lecturer and a research assistant at the CNRS (the French association of research).[4] In 1949 he published his first scientific article, "Sur une partition en cellules associée à une fonction sur une variété". At the same time he wrote a short novel which was only published in Thom (2003). In 1951 he made his PhD in Mathematics. Title: "Espaces fibrés en sphères et carrés de Steenrod" at the University of Strasbourg. From 1952 Thom had a scholarship at the Graduate College in Princeton (USA) and met Claude Chevalley at Columbia University. After 1953 Thom was first a lecturer and later (from 1954–1963) a professor at the Faculty of Sciences in

1 I have used some information contained in the biographical note by Michèle Porte, in Thom, 2003.

2 Preparatory classes specialize students after high-school (bac) and prepare them for the entrance exams to one of the elite colleges ("grandes écoles").

3 The "Agrégation" enables the students to teach at a college.

4 In Thom (1996/2003) he remembers a talk given by Whitney in Strasbourg (1947) on singularities in the plane, a topic which is basic for catastrophe theory.

Strasbourg. In 1957 he was a guest-professor in Bonn (Germany) and in 1958 he received the Fields Medal at the International Congress of Mathematics in Edinburgh. In the sixties he had scientific contacts with a group of mathematicians (Peixoto, Lefschetz) who worked on "structural stability" in the U.S.A. In 1962 he received the "Prix des laboratoires" of the French Academy of Sciences, and from 1963 to 1988 he was a permanent professor at the "Institut des Hautes Études Scientifiques" in Bures-sur-Yvette (Paris). This institution of Advanced Research allows excellent professors their concentration on research at an international level and receives many guests from all around the world.

From 1966 to 1968 René Thom wrote his book "Stabilité structurelle et morphogenèse" which was only published in 1972 (and translated into English in 1975). In 1966 he had published his first article in Theoretical Biology, and in 1967 he had presented the list of seven catastrophes (also called "elementary catastrophes") which are the basis of most applications in biology and the humanities. In 1969 he met Pierre Delattre, a physicist, who founded the Society of Mathematical Biology, which he directed until his death (after that date René Thom took over the direction of this Society). The members met at the Abbey of Solignac in the South of France. The years 1970 to 1976 saw the international success of the applications of catastrophe theory (CT) to many sciences. In 1974 Thom published a series of articles as paperback: "Modèles mathématiques de la morphogenèse" (the edition of 1980 was translated into English, cf. Thom, 1983). While Thom got many international prices and honors, a controversy concerning the application of CT by Christopher Zeeman began (cf. Sussmann and Zahler, 1978). It stopped the rapid application of this type of model building in some fields of applied mathematics. Nevertheless, a large group of scholars from about all disciplines came together at the conference in Cérisy (1982) to discuss René Thom's work (cf. the volume "Logos et théorie des catastrophes" edited by Petitot in 1988).

Figure 1. René Thom in action (as a mathematician).

In 1988 an international mathematical conference in honor of René Thom took place in Paris and Thom presented his new book "Esquisse d'une Sémiophysique". In 1990 he published the book "Apologie du Logos". He received further honors in Poland, France, and Brazil. In 1994 a last congress was dedicated to René Thom's work in the fields of: philosophy, psychology, cognitive sciences, biology, and linguistics. It took place in Paris (cf. Porte, 1994). After 1995 René Thom's health declined and he died on the 25th of October 2002 at Bures-sur-Yvette.

2. The rise and decline of catastrophe theory: Chances for the future

René Thom calls catastrophe theory the *application* of specific mathematical results in the field of differential topology and the theory of singularities.[5] These applications concern rather specific problems in different sciences (from the physics of light, cf. caustics, to biology, neurology, psychology, sociology and linguistics). The underlying mathematics have been further elaborated, e.g., in chaos theory, but they did not become the object of a controversy, i.e., they just followed a general line of further sophistication.

The first phase of CT beginning in 1967 has been dominated by Christopher Zeeman's scientific imagination. He invented a large number of applications reaching from models of the brain to the field of flight/attack in dogs and finally to prison insurrections. At the International Conference of Mathematics in Vancouver (1974), Zeeman had to repeat his presentation for an audience of journalists who received his ideas enthusiastically. As he proposed more and more models (cf. Zeeman, 1977), in which CT was used in different manners (interpreting the internal and external parameters in different ways) his procedure became an object of controversy. It concerned first the partial and often very specific interpretation of the variables in the mathematical model and secondly the fact that no quantitative or statistical test of the hypotheses was given or advocated.

René Thom (in 1996) argues that applied mathematics were at that time (1974) dominated by people using either computers (coming from the information sciences) or statistics (as in economy and psychology). The qualitative nature of Zeeman's models was not acceptable to them. Later, when the qualitative dynamics of Poincaré were again an object of discussion, a better theoretical understanding was reached, but the original impetus or euphoria of catastrophe theoretical modeling was lost or at least weakened. The controversy stopped as soon as 1980, when Smale, whose views had been transported by Sussmann and Zahler (1980), published a review of Zeeman's

5 Remark bis Marc Chaperon (letter): Singularity theory has also been influenced by the observation of unexpected caustics in dimension 3. In this sense, applications have influenced the further development of mathematics.

collection of articles (Zeeman, 1977).[6] Fifteen years after the catastrophe controversy (in 1993), the Dutch mathematician Floris Takens judges:

> Exaggerating somewhat, one can say that where applied mathematics used to be confined to investigate the equations, and their solutions, given by the accepted mathematical models, for the different phenomena, the work of Zeeman showed a much more liberal attitude towards the choice of these models. In this sense he paved the way for abstract parts of mathematics to the applications. (Takens, 1993: 256)

Insofar as Thom was not immediately concerned by the critique of Zeeman's model building in applied mathematics he neither reacted to the critics nor changed his own strategy.

In the domain of mathematics, his proposals were more or less restricted to the field of singularities of a potential, i.e., to gradient dynamics. He agreed that after his early contributions, American and Russian mathematicians (Mather, Milnor, Smale, Arnold) had further advanced the theory of singularities (cf. Thom, 1973/2003: 4), but his primary aim was to further unfold the morphodynamic intuitions he had described in his basic book: "Stabilité structurelle et morphogenèse" (1972). Thom's and Zeeman's proposals had shown that the mathematical results in singularity theory and dynamic systems theory not only had deep historical sources in mathematics (ibidem) but that they were very promising for future applied research.

After the controversy, the tandem Zeeman/Thom separated, and both engaged into different directions, though with a similar intellectual background. Zeeman contributed to the conference on Thom in 1982 (cf. Petitot, 1988: 298–309) with an article on perception ("Sudden Changes of Perception") and he continued his work avoiding however applications in sociology and in other politically sensitive disciplines. In the late seventies, Thom returned to Aristotle and his theory of genres and introduced the terms "prégnance" and "saillance", which make up the heart of his "Semiophysics" (cf. Thom, 1988; Petitot, 1992, and several articles in this volume).

6 After an article in Nature, the replies by mathematicians, physicists, and biologists were unanimously in favour of catastrophe theory (cf. Zeeman, 1993: 268). Smale has at least not made the many mistakes his pupils had shown in the controversy.

3. René Thom's work in linguistics and semiotics

René Thom's interest in linguistics started in 1970, and he mentions (cf. Thom 1996/2003: 29) that his first contacts with linguists had been deceiving, because they rejected the idea of language universals. It was only with the success of Chomsky's and Jakobson's ideas (in the U.S.A. in the sixties; in Europe after 1970) that these linguistic concerns were taken seriously. In Tesnière's grammar, Thom found an application of graph theory to grammar, and thus an intuitive topological model. The basic idea that at the center of the sentence one finds a heavy kernel (the verb), around which lighter elements (nouns and nominal phrases) are turning, invited a gradient-dynamic model (cf. ibidem: 31). This intuition, which Thom also found in Peirce's writings (at a later time in his intellectual evolution), is the starting point of his proposals for a topological semantics. Many of his basic articles on linguistics written between 1968 and 1974 were assembled in the volume of 1974 (reedited in 1980 and translated into English in 1983). The PhD theses of Wildgen (1979) and Petitot (1982) have elaborated these ideas and broadened their field of application (cf. in English Wildgen, 1982, and Petitot, 2003).

Already in 1982, David Ruelle, one of the founders of modern chaos theory, showed the continuity existing between CT and Hopf bifurcations on one side and CT and Feigenbaum fractals on the other (cf. Ruelle, 1988: 103). The so-called "generalized catastrophe" described in Thom (1972) was a cover term for all systems later described in the domain of fractals and chaos theory. However, the rather simple structures of gradient dynamics around one central attractor were not sufficient, insofar as in many domains of application either higher dimensions of attractors (on a circle, a torus, etc.) or chaotic attractors have to be considered (cf. Wildgen and Plath, 2005). The applications of CT therefore (at least after 1982) had to consider further domains of dynamic systems theory. In a collaboration which started in Cérisy-sur-Salle (at the conference on Thom's work in 1982), Laurent Mottron and the present author undertook such an expansion, which led to the book "Dynamische Sprachtheorie" (Wildgen and Mottron, 1987), where the first part dealt also with Prigogine's and Haken's applications of "dissipative systems" and "synergetics" to the humanities. In Petitot (1992: chapter 5) this evolution of the field was also acknowledged.

The intellectual situation in France in the seventies brought about that only the Paris semiotic school (directed by Greimas) became interested in

Thom's work. Petitot and Brandt were both concerned with Greimas and Thom (cf. Petitot 1992, chapter 7, and Brandt, 1992). In the context of the European "Groupe Sigma" and its conferences and at meetings at the Center of Semiotics in Aarhus (Denmark) and Urbino (Italy), the research on dynamic modeling in semiotics and linguistics had been continued, and several authors (Brandt, Petitot, Wildgen) tried to establish a link with west-coast linguistics (Fillmore, Lakoff, Talmy, Langacker). This cooperation was blocked by the rejection of mathematical modeling by this group known as "cognitive semanticists". More recently Per Aage Brandt succeeded in introducing topological thinking into the American school of "Blending and Conceptual Integration" (cf. Brandt, 2004, and the journal "Cognitive Semiotics"). The book series "European Semiotics" (Lang, Bern) in which this volume is published also has a focus on semiotics in the spirit of René Thom.

It would be an interesting topic in the epistemology of science to explain why Chomsky's mathematical linguistics, which was widely ignored or rejected by linguists in the late fifties and early sixties (in Europe until the seventies) could have such an effect in theoretical linguistics and why the even more radical mathematically based hypotheses of Thom (and its later elaborations by linguists) were not successful to the same degree. Chomsky was in fact not a mathematician, although his dissertation (Chomsky, 1955) had the appearance of a mathematical (or logical) treatise. His innovation however fitted the general atmosphere dominated by the information sciences (cf. Thom's remark on the catastrophe controversy 1976–1978). The Brussels school of Prigogine and Haken's series of Applications in Synergetics learned the lesson of the catastrophe controversy and avoided the humanities as domains of application. Only in the late eighties did Haken enter this field with a preference for experimental psychology and neuroscience (cf. Haken and Stadler 1990; Stadler and Kruse 1995; and Haken 1996).

Meanwhile the innovative era of information sciences is closing, insofar as the new technology is a kind of general tool used in all sciences and has lost the character of a generalized metaphor of the human mind. The neurosciences with their window to the mind and genetics took over the role of zones of rapid growth. In a certain sense, modern neuro-dynamics give a more precise and experimentally more sophisticated view on the mind than Zeeman's articles did in the late sixties. But the general prospect to develop dynamic models of the human mind, of thought and language, persists. Thus, the endeavor of Thom and Zeeman, and of their colleagues in biology, psychology, and linguistics, was not lost. Their enterprise was just a long journey and in the eighties and nineties the scientific community was not eager to wait until they could come nearer to their goal.

It cannot be the intention of this introductory chapter to make guesses for the future, but one can already feel the overall changes that reduce the charm of mechanical (computer-based) models in the humanities. This can either trigger an (impossible) return to folk-models of the mind and language (cf. my critic of cognitive semantics in Wildgen, 1994: chapter 1 and in Wildgen, 2008: chapter 8 in German) or lead to a renaissance of dynamic model building in the humanities (surely with better empirical data and more advanced conceptual tools). The general chance of this direction is that it introduces new mathematical models that have been successfully applied in the natural sciences and uses them to reshape traditional questions and strategies for their solution in the human sciences.

4. An overview of the contributions in this volume

The contributions to this volume are elaborations of papers given at a conference with the same title organized by the editors in Urbino (Italy), 18th to 20th of July 2005. Some of the contributions which were in Italian and not specifically linked to René Thom's work were not considered, other contributions were invited after the conference.

I all cases, the work of René Thom and in many cases his personal influence had been crucial for the contributors in some phase of their academic life or throughout. Therefore many of contributions are a kind of testimony (cf. the title of Marc Chaperon's paper) of the impact René Thom had in the field of semiotics and linguistics. The majority of papers are in English or have been translated into English; they form the first part of the volume. Two papers are in French, and they stand in the second part. As Thom's oeuvre is written in French and many of his books and papers have been translated into English, such a bilingual volume seems to be adequate.

The first chapter by Marc Chaperon is written from a mathematical perspective and describes the intellectual itinerary of René Thom in the light of one of his students. He also comments the controversy about the applications of catastrophe theory (mainly those of Christopher Zeeman) in the late seventies.

The second chapter by Svend Østergaard concerns the recognition of forms and its contribution to cognitive (and pragmatic) realism. Singularity theory is shown to be decisive for a theory of human (and animal) form recognition.

The multiple sources of Thom's semiotics, mainly his "biosemiotics" are enumerated and commented in the chapter by Peer F. Bundgaard and Frederic Stjernfelt. They refer to ideas taken from: Waddington, Uexküll, Saussure, Jakobson, Tesnière (and Aristotle), Köhler, Pike, Greimas, Greenberg, and Durand. Thus they can show the multiple roots of Thom's semiotics and his capacity to integrate a great variety of proposals in a multidisciplinary field and to put forward a mathematically founded theoretical architecture which could bring all these endeavors together.

The search for relevant intellectual sources is continued in the chapter by Wolfgang Wildgen, who concentrates on the notions of "saillance" and "prégnance" and the proximity they show to concepts proposed by Pavlov, Lorenz, Uexküll, Gibson, and Cassirer ("symbolische Prägnanz"). Cassirer's objections against a strict parallelism between notions in modern geometry and the dynamics of human perception are considered in the light of Thom's proposals. The notion of "diffusion de prégnance" is reconsidered in the context of evolution and questions about the origin of grammar (and syntax).

The chapter by Isabel Marcos applies basic proposals of Thom to structural geography and the morphogenesis of towns. She exemplifies the type of morphogenetic analysis based on the historical evolution of Lisbon and she tries to find some morphological "universals" of urban evolution.

Angel Lopez Garcia gives a rather critical reading of Thom`s proposals related to verbal semantics and the structure of basic propositions. He compares Thom's specific analysis of verbs with the tradition of structuralism (Hjelmslev, Jakobson, Halliday, Chomsky) and the models: "Liminar Grammar" and "Topological Linguistics" proposed by himself. Finally, the question of the neurological plausibility of these proposals is put forward.

The chapter by Jean Petitot (in French) begins with a "testimony" of Thom's intellectual development since the end of the sixties and addresses the central question in Thom's linguistic thinking: How can predication be conceived in a "morphological" (topologico-dynamic) perspective, and how is it related to the classical "logico-centric" models. Jean Petitot starts from Husserl's "Erfahrung und Urteil" and introduces specific mathematical concepts like "faisceaux" and "topoï". They allow for an elaboration of the logical treatment of predication as it was put forward by Kripke and others by adding a "morphological" dimension.

The chapter by Per Aa. Brandt (in French) returns to a less formal treatment of "prégnance" and meaning in an elaborated structural design based on Saussure's dichotomy of "signifiant" and "signifié". He expands the idea of "diffusion de prégnance" towards a model of double "flux" between "signifiant" and "signifié".

References

Arnold, V.I. (1986). *Catastrophe Theory*, Heidelberg, Springer.

Brandt, Per Aage (1992). La charpente modale du sens. Pour une sémio-linguistique morphogénétique et dynamique, Aarhus University Press and John Benjamins, Aarhus/Amsterdam/Philadelphia.

Chomsky, Noam (1955). *The Logical Structure of Linguistic Theory*, manuscript electronic copy; partial publication, Plenum Press, New York, 1975.

Haken, Herman and Michael Stadler (eds.) (1990). *Synergetics of Cognition. Proceedings of the International Symposium at Schloß Elmau*, 1989, Springer, Berlin.

Haken, Hermann (1996). *Principles of Brain Functioning. A Synergetic Approach to Brain Activity, Behavior and Cognition*, Springer, Berlin.

Nicolis, Grégoire and Ilya Prigogine (1989). *Exploring Complexity. An Introduction*, Freeman, New York.

Peirce, Charles Sanders (1931–35). *Collected Papers of Charles Sanders Peirce* (ed. by Charles Hartshorne and Paul Weiss), Harvard U.P., Cambridge (Mass.), vol. I–VI (cited with vol. and §).

Petitot-Cocorda, Jean (1985). *Morphogenèse du sens*, Presses Universitaires de France, Paris (English translation: Petitot 2003).

Petitot-Cocorda, Jean (ed.) (1988). *Logos et Théorie des Catastrophes*, Patiño, Genf.

Petitot-Cocorda, Jean (1992). *Physique du sens. De la théorie des singularités aux structures sémio-narratives*, Éditions du CNRS, Paris.

Petitot, Jean (2003). *Morphogenesis of Meaning*, translated by F. Manjali, Lang, Bern.

Porte, Michèle (éd.) (1994). *Passion des formes*, 2 vol., Fontenay-Saint Cloud, E.N.S. Éditions.

Ruelle, David (1988). "Idéalisation en physique", in: Petitot (1988: 99–105).

Stadler, Michael and Peter Kruse (eds.) (1995). *Ambiguity in Mind and Nature*, Springer, Berlin.

Sussmann, H.J. and R.S. Zahler (1978). "Catastrophe Theory as Applied to the Social and Biological Sciences: A Critique", in: *Synthese* 37, 117–216.

Takens, Floris (1993). "The Work of Professor Sir Christopher Zeeman FRS", in: *Niew Archief voor Wiskunde*, 11 (3): 251–256.

Tesnière, Lucien (1959). *Éléments de syntaxe structurale*, Klinksieck, Paris.

Thom, René (1972/1977). *Stabilité structurelle et morphogenèse. Essai d'une théorie générale des modèles* (2nd edition, revised and augmented 1977), InterEditions, Paris (citations are based on this edition).

Thom, René (1974a). *Modèles mathématiques de la morphogenèse*, Paris, Union Générale d'Editions (2e édition, Bourgois, Paris, 1980).

Thom, René (1975). *Structural Stability and Morphogenesis*, Benjamin, Reading (translation with some omissions from the French original: Stabilité structurelle et morphogenèse 1972).

Thom, René (1988). *Esquisse d'une sémiophysique. Physique aristotélicienne et théorie des catastrophes*, InterEditions, Paris (translation into English: 1990. Semiophysics: A Sketch, Addison-Wesley, Redwood City, Calif.).

Thom, René (1990). *Apologie du Logos*, Hachette, Paris.

Thom, René (1996/2003). "Document 4.2.: Graphes-Signaux-Prégnances", in: Thom (2003: Inédits 1997: 29–36).

Thom, René (1997/2003). "Document 1: Mémoire de la théorie des catastrophes", in: Thom (2003: Inédits 1997: 1–28).

Thom, René (2003). *Oeuvres complètes*, CD-Rom, Institut des Hautes Études Scientifiques, Bures-sur-Yvette, Paris.

Wildgen, Wolfgang (1979). *Verständigungsdynamik. Bausteine für ein dynamisches Sprachmodell*, thesis of Habilitation, University of Regensburg; electronic copy: <http://elib.suub.uni-bremen.de/ip/docs/00010402.pdf>.

Wildgen, Wolfgang (1982). *Catastrophe Theoretic Semantics. An Elaboration and Application of René Thom's Theory*, Benjamins, Amsterdam.

Wildgen, Wolfgang (1985). *Archetypensemantik. Grundlagen für eine dynamische Semantik auf der Basis der Katastrophentheorie*, Narr, Tübingen.

Wildgen, Wolfgang (1994). *Process, Image, and Meaning. A Realistic Model of the Meanings of Sentences and Narrative Texts*, Benjamins, Amsterdam.

Wolfgang Wildgen und Laurent Mottron (1987). *Dynamische Sprachtheorie. Sprachbeschreibung und Spracherklärung nach den Prinzipien der Selbstorganisation und der Morphogenese*, Brockmeyer, Bochum.

Zeeman, Christopher (1977). *Catastrophe Theory: Selected Papers 1972–1977*, Addison-Wesley, Cambridge (Mass.).

Zeeman, Christopher (1992). "Controversy in Science: on the Ideas of Daniel Bernoulli and René Thom", in: *Niew Archief voor Wiskunde*, 11 (3): 257–282).

Part One
Chapters in English

Catastrophes. A testimony♣

MARC CHAPERON

1. Encounter

When I first met Thom, at the end of 1973, he was just fifty. At the *École Normale Supérieure*, where I had been admitted four years earlier with more musical projects than scientific ones, I had been astonished that our mathematics teachers would give us "the good viewpoint" without telling us on what.[1] I had then sworn to myself that I would have nothing to do with those people, without realising that only laziness prevented me from fetching the (considerable) unspoken part of their message.

During my stay at the school rue d'Ulm, gradually renouncing my adolescent dreams, I had engaged into not very intoxicating "applied" mathematics. Then, I read *Stabilité structurelle et morphogenèse* (Thom 1972a)[2], of which I did not understand much[3], apart from the facts that it contained magnificent mathematics and that I was in the presence of a rare phenomenon: a mathematician who thinks. Then, catastrophe! The engineering school where I should have had my first teaching position did not manage to ask for me in time, so that I had to spend one academic year with the very heavy responsibility of a *mathématiques spéciales* class.[4]

The disaster played an amazingly positive role in my life. When Michel Hervé, then co-director of the *École Normale Supérieure*, tried to help me, I

♣ I wish to thank the organisers of this worthy tribute to the memory of René Thom. It is essential that we, coming from the rather various domains which composed his world, should put in common what we have learned and transmit it. Indeed, History teaches us that this is the only way to pass knowledge from one generation to the next. This might be what Thom meant when he quoted Max Planck: "Science progresses by the death of the old".

1 Paraphrasing Victor Hugo's "Do not drop music along my verses", the more radical members of Bourbaki would proclaim: "Do not drop sense along my mathematics".

2 References between brackets are listed at the end of the article.

3 After some time, I guessed that (almost?) everyone was in the same situation.

4 Where supposedly gifted students are prepared to the competition for "great" (and lesser) scientific schools.

confessed that, had I known of it earlier, I would have tried and followed the way opened by Thom. He replied: "That would make something of a difference. Go and see him!" and succeeded in overcoming my strong feeling of indignity.[5] This is how Thom accepted me as his student, saving me both from pure teaching and from scientific boredom.

He called me "Monsieur le professeur de spéciale" and was astounded by my ignorance.[6] I owe him great moments such as a crystal-clear course in homotopy, improvised for me in his office at the I.H.E.S. He advised me to start with a thesis in pure mathematics ("much easier than applied mathematics"[7]) and had me read his article *Sur le cut-locus d'une variété plongée*[8] (Thom 1972b).

Twenty years later, I realised that he must have been wishing me to attack the Poincaré conjecture from that side, but I completely missed his point at the moment: I was too ignorant and too much marked by the ambient ideology of doing everything in maximal generality. Having thus considerably disappointed him, I imagine, I found my road to Damascus in his article (Thom 1972c), a wonderful introduction to singularities and contact geometry. Thom had been delicate enough to forget a curve in one of his statements, which enabled me to give my first mathematical talk at his seminar during the fall of 1974.

There were two Thom seminars at the I.H.E.S., one on mathematics and the other on catastrophes.

---

5 A feeling which never left me in my relationship with Thom.

6 I had chosen mathematics because it was easy to succeed there with only the famous "good viewpoint" in one's pocket. While exclaiming on my ignorance, Thom kindly told me that he had never learned anything without a problem to solve.

7 I do not resist the temptation to quote the sequel: "Since Von Neumann, has there been any great applied mathematician besides Nash? Kolmogorov, Moser and Arnold are probably greater as pure mathematicians."

8 The "cut locus" is the set of those points in the ambient Riemannian manifold whose distance to the given submanifold is reached at several points (provided they are counted with their multiplicities).

2. Thom's mathematical seminar

It used to take place on Monday afternoons, switching to Friday afternoons after Thom was elected at the Academy of Sciences. Most of the speakers were first rank mathematicians, who explained without unnecessary complication what they had been doing in one of Thom's numerous domains of competence. He had the art of expressing in one sentence the essence of a lecture, which he would sometimes reveal to the lecturer himself. It was fascinating and incredibly instructive. I learned there most of what I know in mathematics. Much to my surprise, almost none of my fellow students of the École Normale Supérieure was attending it: instead, the competent (?) authorities had recommended them dreadfully formalist seminars of which they would understand practically nothing, feeling nevertheless very proud of the little they could grasp.

3. The climate of the time

Indeed, my generation loved what was beyond understanding, considering obscurity a token of quality. The French society was more than ever regarding art and knowledge as sacred, due to the legacy of Baudelaire, Mallarmé or Valéry, of course, but also because the communist party was exerting a strong influence on criticism and, more generally, on the press.[9] Some would ceaselessly take advantage of the situation: on the pretext that many geniuses have been misunderstood, more or less talented individuals would produce an incomprehensible work and become subsidized geniuses of the Republic.

This taste for "transcendence" and its corollary, the abdication (sometimes) of critical sense, often had quite regrettable consequences. Did not the Maoists, right in the middle of the Cultural Revolution, swallow hook, line,

9 Notwithstanding the accent put on culture in the Soviet Union, which produced one of the most educated populations in the world, it is likely that, in those days of cold war, the Soviet propaganda was flattering the artists and intellectuals of the West only to use them better. Remember Aldous Huxley's despair when the caricatures of a Soviet satirical magazine made him discover, after the forces of the Axe entered the war, that his beloved Peace Movement had been manipulated by Stalin to prevent the Allies from getting ready.

and sinker all the fibs of a cruel and irresponsible government? Looking for new Gospels, we sometimes made totally wrong choices. What a waste of time, intelligence and good will!

4. Publication of *Stabilité structurelle et morphogenèse*

Unfortunately, I am not sure that such demission of the mind was foreign to the dazzling media success of Thom's ideas. Of course, as a mathematician, he was luminous, combining an outstanding abstraction power with a kind of "country common sense" which would keep him close to the essential.[10] The problem is that such a concise, often elliptic thought, whose comprehension demanded already a huge effort from mathematicians, was clearly not accessible to the public.

Obviously, the latter could not understand *Stabilité structurelle et morphogenèse* despite Thom's efforts: its mathematical content was too omnipresent and, even to people like myself, trained to grasp a good part of it, the work remained quite mysterious as a whole. By the way, Thom's literary culture[11] may have made him share Jacques Lacan's belief that an obscure text is enriched with its many potential interpretations.

In spite of this very large part of mystery or because of it[12], the book was a dazzling success. One must say that the British mathematician E. Christopher Zeeman had not been skimping on propaganda, urging the *mass media* to proclaim: "The new Newton is French"[13]! The same astounding success was met by a set of lectures and articles by Thom, *Modèles mathématiques de la morphogenèse,* which appeared in a popular pocket collection soon afterwards (Zeemann 1965).

Catastrophe theory had become a *fashion*, a necessary reference (reverence?) in the discourse of the subsidized vanguard in the same way as Marxism and psychoanalysis. Even old Dali made it one of the dominant themes

10 "The questions that must be asked and not those that can be asked", to quote Poincaré.

11 Rather Mallarméan, judging by the quotations in his book: Valéry, Huysmans, the pre-Socratic …

12 See section 3 of this article.

13 Title of the weekly magazine L'express in October 1974.

of his last works. Someone went so far as to commit a book on the subject bearing this edifying subtitle: *How to avoid personal disaster*!

5. The Saturday seminar

The second Thom seminar, which took place on Saturday mornings, did not have the same public as the first, even though mathematicians like Jean Cerf, Alain Chenciner, Larry Siebenmann or (when in Paris) Jorge Sotomayor attended it regularly. I did not feel at ease there, due to the presence of slightly too unconditional faithful and of careerists waiting for the last intellectual fashion to help their ascension in the academic world. My beginner status was forbidding me to say anything, but it seemed to me that *this was not right.* I had understood enough of catastrophe theory to know that it was a (fruitful) viewpoint and a (fascinating) programme rather than a complete scientific theory. There, most of the questioning came from Thom himself.

I am not sure that he benefited from the situation: whereas his mathematical successes had often been obtained by his differentiating from a nevertheless exceptional entourage[14], he did not meet much resistance among his usual interlocutors in the adventure of catastrophes. Or a sly resistance: I remember with horror the carnivorous smile of a supposedly philosophical skinhead who had succeeded in showing that the master might not know every corner of Kant …

Anyway, there is no doubt that the Saturday seminar reflected Thom's interests at that moment of his life. It may have been a refuge against the reprobation of many French mathematicians, a reprobation based on envy (think! He could be seen on television) rather than on a serious analysis of his work.

It must be said that Thom had a pronounced taste for provocation, with the same result as for Vladimir Arnol'd a little later, and for the same reasons: not realising how much he weighed in the opinion of the scientific community, he wanted to play with them, with the result that they felt savagely aggressed. This certainly did not contribute to the diffusion of his ideas, espe-

14 For example, in the beginning, he resisted Henri Cartan's desire to have him work in complex analysis when he realised that the problems posed to him in the holomorphic framework were unsolved in topology.

cially among biologists; and the reprobation of the French mathematicians partly came from their fear that their community should pay for his excesses.

6. Catastrophes according to Zeeman

To a large extent, the history of catastrophe theory coincides with that of the relationship between Thom and Zeeman, whose article *Topology of the brain* (Zeeman 1977) played an extremely important role in the elaboration of the models proposed in *Stabilité structurelle et morphogénèse*. Zeeman demonstrated almost Chinese modesty on the subject, saying in substance that his bad article had contributed to the birth of a great theory.

It is probably under this influence that Thom, in his book, systematically took space-time as a *parameter space* for the dynamical (or static) systems under study. This explains why he gave so much importance to the *elementary catastrophes*, which are the only local accidents that can affect a family of potentials depending on at most four parameters and "generic" in the sense that it has nothing special.[15]

This idea of space-time as a set of "control variables" for a phenomenon described by other "internal variables" looked frankly bizarre to many scientists, who bore in mind cases where things happen differently. However, there are several important examples of it. In Hamilton-Jacobi theory, for instance, the solutions $u(t,x)$ of a partial differential equation $\partial_t u + H(t,x,u,\partial_x u) = 0$ appear under the form $u(t,x) = \min_v S(t,x,v)$, where S is a differentiable function on the product of space-time by a space of parameters v; as the minimum of S is taken with respect to v, the singularities of u will indeed be obtained by viewing S as a family with parameter (t,x) of functions of the variable v. This can be found[16] in the fourth chapter, "Cinématique des formes, catastrophes" of *Stabilité structurelle et morphogénèse*.

---

15 The tool enabling one to formulate the notion correctly is Thom's transversality lemma, one of his major contributions to mathematics.

16 In the case of wave fronts. In geometrical optics, for example, one rather considers as a solution the graph of the multi-valued function which to (t,x) associates the set of critical values of S(t,x,v) viewed as a function of v: as there are very few photons in light rays, the latter can cross each other without any problem. The solution u is obtained by choosing the lowest point over each (t,x).

Zeeman proposed many applications of catastrophe theory[17] where the space of parameters was not space-time. Their spirit was more pragmatic than Thom's; his ideas were often applied to domains which were comparatively devoid of risk, being closer to physics and therefore strongly mathematical already[18]. Thom was not much interested, for he thought that "when you know where you are going, you do not go very far"!

7. The catastrophe of catastrophe theory

Between 1972 and 1978, the craze for catastrophes led to the funding of numerous research programs throughout the world. At that point, the (good) mathematician Héctor Sussman, a former propagandist of catastrophes, writes with Raphael Zahler a very severe critique of their applications in the human sciences and biology (Sussmann and Zahler 1978). After the frequently absurd fashion comes another one[19], which consists in denigrating what had been praised to the skies.

The rats then leave the sinking ship: mathematicians who were paid to study catastrophes hurriedly take their distances with respect to Thom and Zeeman, sometimes without the slightest elegance–I remember an article where two scoundrels took advantage of a blunder of Zeeman to claim, against all truth, that "catastrophe theory could not yield their results".[20]

If such U-turns left Thom indifferent, the criticisms (Smale 1978)[21] of the American mathematician Stephen Smale, whom he admired and had much influenced, came as a hard blow. I remember that, around that time,

---

17 Considering chiefly "elementary catastrophes", i.e. singularities of differentiable functions (Thom's "static model"). An anthology of articles by Zeeman is (Zeeman 1977).

18 There are exceptions, such as his 1975 work on anorexia. It appears that on this occasion Zeeman collaborated with a therapist who somewhat lacked seriousness. Nevertheless, such models remain far preferable to the painful puns of the Lacanian School, as they save psychoanalysis from the tyranny of language.

19 See for example Gina Kolata (1977), soon relayed by the newspapers (*New York Times*, 19 November 1977).

20 Some resisted: another north-American mathematician wrote in substance: "One could say that what we have just seen looks precisely like a cusp catastrophe if *catastrophe* had not become a dirty word."

21 The pretext was the review of Zeeman's already mentioned anthology (Zeeman 1977).

while driving him from Bures to Paris, I was unfortunate enough to emit the idea that there might remain some work to do on his theory; immediate reaction: "So, you are like Smale: for you, catastrophe theory does not exist!"

Smale's criticisms deserve some discussion for, apart from their psychological impact on Thom, they pose the problem in clear, relatively moderate terms. In the introduction already, the rhetoric precaution to attribute Smale's negative opinion to his "conservatism in science" is followed by a nourished salve: "Some of the mathematics underlying CT, especially transversality, and singularities of maps, has played a constructive role in outside disciplines, and is destined to play an ever increasing role. On the other hand I feel that CT itself has limited substance, great pretension and that catastrophe theorists have created a false picture in the mathematical community and the public as to the power of CT to solve problems in the social and natural sciences."

The initial tribute to transversality and singularities of maps is of course paid to Thom as a great mathematician. To a large extent, the essential reproach ("a false picture") is justified, but I would say that Thom has very little to do with it whereas Zeeman is "responsible but not guilty", to quote the word of a French politician.

In the British good society, submitted to an almost exhaustive etiquette, eccentricity is not only tolerated as a security valve, but also extremely well considered. When Zeeman praises catastrophes, even though he does go somewhat too far, this exaggeration is not dishonest at all: he behaves according to the social codes of the world he belongs to. No more than Thom is he striving for vulgar material success: it is in the kingdom of ideas that both intend to leave their trace; the European academic tradition, in that blessed period, is still protected from the devastating power of money.

Unfortunately, such is not the case elsewhere, where the allocation of comfortable research credits imposes a relatively quick return on investments. Inside that system, Smale's criticism is obviously pertinent: in five or six years, one does not have enough time to impose as normal practice a "general theory of models" which has been only sketched[22] and contrasts sharply with former habits. One can at most, as Zeeman, give examples intended to convince the scientific community to go further, without pretending to catch the moon at once.

That catastrophe theory has "limited substance" seems more dubious to me. Even as a pure mathematician, Thom was a visionary, more interested in

22 See the second part of the title of *Stabilité structurelle et morphogénèse* (Thom 1972a).

producing new ideas than in working them out technically. Even though the models he proposed in *Stabilité structurelle et morphogenèse* may not have always been relevant in detail, the very strong underlying ideas cast a new light on the notion of a mathematical model, justifying Zeeman's dithyrambs, I believe. We shall now make this more precise by examining Smale's third criticism.

8. "Great pretension"?

Thom and Zeeman found themselves in the situation of who has discovered a gold mine and cannot extract from it enough nuggets to convince investors to exploit it seriously. They may not have been very skilful when it came to handling a pickaxe, for they were definitely not applied mathematicians. Here, Smale's reproaches (Smale 1978) are easy and, to a large extent, unfair: "good mathematical models are not generated by mathematicians throwing models to sociologists, biologists, etc. for the latter to pick up and develop. [...] Good mathematical models don't start with the mathematics, but with a deep study of certain natural phenomena".

Now, even the most beautiful girl in the world can give only what she has: what Thom and Zeeman had at their disposal was a mathematical breakthrough whose significance for applications seemed clear to them. Therefore, they tried to convince the scientific world to use it. This attempt for seduction was not always adroit, to say the least. Zeeman, having shown so much optimism during his propaganda campaign, was feeling compelled to prove quickly what he had claimed, bypassing collaborations with other disciplines which would have reinforced his models–hence rejection by the specialists he had had no time to consult. And Thom's irony in front of this rejection was not bound to settle the problem of the missing relays.

Another aspect of his psychology was involved, on which Georges Lochak puts his finger in his homage (Lochak 2002) to Thom: "He told me that, when he was admitted to the École Normale, in mathematics of course, he went to see the director and said that he would rather do philosophy." Thirty years later, with his glorious mathematical record, Thom wanted to realise the dream of his youth[23] and obviously considered his mathematical work as finished. This also provides an explanation for his violent reaction during our little trip in my automobile: judging that he had said or thought the essential in the domain of catastrophes[24], he was not keen to admit that what he considered an achievement might be taken by others as a starting point. That attitude of course did not help those who, from Floris Takens to the author of these lines to Alain Chenciner or Robert Roussarie, wished to continue his reflections on singularities of dynamical systems, for example. Vladimir Arnol'd alone, in Moscow, was strong enough to stick to it despite the reflux of fashion and the distant look cast by Thom on his followers.[25]

Anyway, Smale's idea of "good mathematical models" does illustrate his "conservatism in science", as it negates Thom's essential contribution: provide the modelling of complex situations with a methodology founded on the search for some universal morphologies – the catastrophes. The idea is not entirely new – as often, it can already be found in Poincaré – but here it becomes much more explicit and has not stopped bearing fruit. Whatever Smale may claim (Thom too: "When you know where you are going, [...]"), unless one knows what to look for, there is little chance of finding it. Of course, one may risk to incur the reproach addressed by the French novelist Julien Gracq to the structuralists: "As they think they have found a key, they will not rest until they have given your work the shape of a keyhole!" but the risk decreases when the "stock" of models increases. Some of the most remarkable recent work in biology and, even, astronomy, widely proves Thom right against Smale.

23 Had he ever forgotten it, by the way? As Lochak excellently puts it, "everything he was interested in was meant to serve one cause: that of his vision of the world", so that his *mathematical* work is far more philosophical than most philosophical works.

24 To a large extent, for him, "catastrophe theory" was the sum of all his reflections – and he had reflected much – rather than the content of such or such text, *Stabilité structurelle et morphogénèse* included. He therefore took Smale's criticisms as a personal attack.

25 I believe this distance to have caused the grievance expressed rather aggressively by Arnol'd (1984) in his wonderful little book on Catastrophe Theory .

References

Arnol'd, V. I. (1984). *Catastrophe Theory*, Springer, Berlin (third corrected and expanded edition, 2002).

Kolata, Gina (1977). "Catastrophe theory: The Emperor Has No Clothes", in: *Science*, April 15.

Lochak, G. (2002). René Thom (1923–2002), in: *Annales de la Fondation Louis de Broglie*, Volume 27 n° 4.

Smale, S. (1978). Review: E. C. Zeeman, Catastrophe theory: Selected papers, 1972–1977, in: *Bulletin of the American Mathematical Society* 84, 6: 1360–1368.

Sussman, H. and Zahler, R. (1978). "Catastrophe Theory as Applied to the Social and Biological Sciences: A Critique", in: *Synthese* 37, 2: 117–216,.

Thom, R. (1972a). *Stabilité structurelle et morphogenèse. Essai d'une théorie générale des modèle,* Benjamin, Reading, MA,. Deuxième édition revue, corrigée et augmentée, Ediscience, Paris, 1977. *English* translation by D. H. Fowler. *Structural stability and Morphogenesis. An outline of a general theory of models*, Benjamin, Reading, MA, 1975.

Thom, R. (1972b). "Sur le cut-locus d'une variété plongée", in: *J. Differ. Geom.*, 6: 577–586.

Thom, R. (1972c). "Sur les équations différentielles multiformes et leurs intégrales singulières", in: *Boletim da Sociedade Brasileira de Matemàticas*, 3, p. 1–11,.

Thom, R. (1974). *Modèles mathématiques de la morphogenèse*. UGE, Paris (collection 10/18,. Second corrected and commented edition, 1980).

Zeeman, E.C. (1965). Topology of the brain, in*: Mathematics and Computer Science in Biology and Medicine*, Medical Research Council.

Zeeman, E.C. (1977). *Catastrophe Theory: Selected Papers*, 1972–1977, Addison-Wesley, New York.

René Thom: The Recognition of Forms. An Apologia for Realism

SVEND ØSTERGAARD

In a few short statements I will present René Thom's philosophical commitments. As is well known, René Thom was the discoverer of catastrophe theory, but more generally, we can say that he made substantial contributions to the theory of dynamical systems (mathematical analysis) and to the theory of differential manifolds (topology). These fields influenced his conception of mathematics: he was very much against set theory, partly because it did not contribute to any understanding of the perceptual world and partly because it led to incomprehensible paradoxes. He was not a formalist, but he accepted of course formal representations in mathematics insofar as they led to insights in geometry, topology and dynamics. So, mathematics is not primarily a formal system, but a method to obtain realistic representations of space and dynamic processes. One can perhaps say that he reverted to Newton's understanding of mathematics: The mathematical laws were predicates of the objects, i.e. not anything imposed from outside. If mathematics describes structural properties of space/dynamics on the one hand, then on the other it also describes the structural conditions for the subject's recognition of the external world. René Thom almost favors a mirroring effect where the dynamics of the external world correlates with a dynamics of the human brain, and this correlation makes recognition possible. This is a position which underlies a philosophy that contains the following assumptions.

The continuum

The only geometric entities that exist in the physical world are closed, bounded forms with maximal dimension, i.e., forms in 3 dimensions. This intuition stems from Aristotle, according to whom the point does not exist in isolation. The point only exists as a virtual structure, in principle the point has to be considered as a small oscillator that spans an infinitesimal line segment. The same can be said about the line: The line only exists insofar as it generates a surface, which only exists insofar as it generates a volume. So, only volumes exist and the continuum of space becomes a basic aspect of the ontology of Thom. In this regard, Thom is referring to Aristotle's Physics, where he finds the first attempt to describe the world from the perspective of the continuum (cf. Thom 1998). In this way he wants to avoid the problems inherent in numbers and especially the problems that occur when we identify numbers with points on a line.[1] Of course, René Thom will not refrain from using the advantages of representing points by means of numbers. In geometry this goes back to Descartes' analytic method whereby forms are represented as solutions to equations. This approach has shown its immense power in the history of mathematics, but accepting the practical solution of representing forms as a collection of points that appear as solutions to equations does not entail any ontological commitment. Thom's assumption of the ontological primacy of the continuum leads to a rejection of the point as an existing entity in the world; it only exists as a practical concept.

Saliency and pregnancy

The only manifolds that have ontological reality are forms in space. Thom calls these forms *saliencies*. Forms appear, disappear, merge, interact through contact etc. Through this interaction the forms influence each other: one form makes another move or one form changes another. Thom subsumes this dynamic propensity under the name of *pregnancy*. In other words: the salient forms contain pregnancies that might influence other forms, maybe even be transported from one form to another, as when a billiard ball hits another ball and sets it into motion. The philosophical commitment of this is that dynamics is something that is inherent in the forms of the world. This represents a contradiction to for instance some aspects of modern physics according to which dynamics is equivalent with fields that exist independently of the actual object subjected to these fields.

1 In passing, we can note that in this respect Thom is in opposition to other trends in French philosophy which would rather consider the enigmatic aspects of set theory as the foundation of ontology; this is for instance demonstrated in the work of Allan Badiou.

According to Thom it is for any ontology necessary that it is possible to distinguish between saliencies that can be localized in a geometric space and pregnancies that are not localizable entities, but which are parts of the salient forms. If a theory only operates with saliencies, or if it separates the saliencies from the pregnancies, it becomes inexplicable how one form can influence another; causality becomes an instance of magic. So it is in the recognition of dynamic processes that human thought has to operate with pregnancies. An example: in classical physics the salient forms are the material objects often idealized to points, the pregnancies can be the momentum of velocity which can be transported from one form to another. For instance, if a billiard ball sets another billiard ball into motion, to understand this as causal it is necessary to assume that "something" has been transmitted from the first ball to the next. In Aristotelian physics this "something" is force, but in the classical physics it is momentum of velocity. In biology the salient forms are the biological organisms and the pregnancies are hunger, fear, sexual desire etc. which regulate the dynamics of the organisms.

Structural stability

The salient forms in space and their dynamic interaction constitute the phenomenal world. René Thom's mathematics is developed under the assumption that the structural properties of the phenomenal world correlate directly with properties of mathematical structures, mainly differential manifolds. In this approach the relation between empirical data and theory is different from classical physics where the structural relations are dependent on experiments and formal models that explain the experimental data. In the qualitative physics the empirical knowledge is not dependent on experiments, but is directly accessible to the mind. This means that the mathematical theory is about why and how the world is experienced in the way it is, and why this experience is unproblematic. This does not mean that qualitative physics is elementary; in a way the physics is simple in the experience of the subject, because it is directly experienced, but the formalism explaining why it is so is very complex, whereas in classical physics the formalism is in some cases simple, think of Newton's second law or Einstein's equation for energy $E = mc^2$, but the physics on the other hand is very complex, depending on non-visible entities. One of the reasons for the complexity of the qualitative physics is that its goal is to uncover the schematic relations in the phenomenal world that make it possible for the mind to ascribe meaning to the world. One of the conditions for this is that the phenomenal world is not chaotic, but, on the contrary, *structurally stable*. The notion of structural stability is therefore an important part of Thom's ontology, and one of his main contributions in mathematics and semiotics is to present precise determinations of the forms in which structural stability can occur.

Structural stability refers both to forms in space and processes in time. If we take a form in space – for instance, a vase – then it is bounded in relation to the surrounding space. This boundary is in a way a critical border which determines the identity of the vase; however, it is structurally stable since it is resistant to small dynamic influences; I can for instance tap on the vase with a pencil without it being broken. If the boundary is not resistant to minimal influences, then we have a phenomenon that has to be described in time. For instance, a ring of smoke which first appears as a form (a ring), but then it is gradually dissolving and the salient form disappears while its pregnancy influences the surrounding (the smoke might color the wall paper etc.).[2]

More generally, the recognition of an object is not only dependent on its contour. It might also be that the object consists of qualitatively different zones. The recognition of the object is therefore dependent on the recognition of a set of critical boundaries along which different qualities are in conflict. To exemplify, imagine a self-portrait by Van Gogh. Imagine that the difference between background and foreground is marked by the direction of the stroke; for instance, vertical stroke in the portrait and horizontal stroke in the background. The directions of the stroke – vertical vs. horizontal – are thus two qualities with each their zone and the border between the zones constitutes a critical boundary where the qualities are in conflict. This determines the perceptual experience of looking at the portrait, and one can easily lift this to a semiotic interpretation of the portrait illustrating the precarious conflict between the subject and its environment.

The identity of a non-chaotic process is similarly determined by critical points where the system at a phenomenological level changes from one state to another. For instance, if we heat ice it changes to water and later to steam. The critical points of the process are normally structurally stable in the sense that small changes of the environment don't change the qualitative features

2 It must be said that this distinction between forms and processes is crucially dependent on humans' phenomenological experience of time, since in a grand time scale also the forms are subject to changes.

of the process; for instance, the above mentioned process of heating ice does not change qualitatively if we make small changes of the atmospheric pressure.

The topologic structures that describe forms in space and processes in time are, according to Thom, a priori conditions for what qualitative forms it is possible to recognize at all. One has to understand this correctly, it is of course not the case that all forms in the world can be found in mathematics, but every configuration of critical boundaries/points that determines the identity of a form or a process can be categorized in differential topology. Since Thom is only interested in the phenomenal world, i.e. processes in space/time, the mathematical investigation can be limited to four dimensions, and this leads to a finite list of possible catastrophic sets and a finite list of possible processes. The transformations and deformations of this list are infinite, though. If a verb refers to a process in space, then the schematic content of the verb should in principle be identified with a trajectory in a differential manifold, cutting a finite set of singular points. For instance, "he entered the room" is a process with only one critical point, namely the crossing of the line that separates outside from inside. Such a general categorization is presented in Thom (1972 and 1980).

Perception

Let me summarize: The critical areas in space-time where qualities are in conflict constitute a set of singularities in the four dimensional space-time. This set of singularities is a topological-geometric structure that can be classified in differential topology. The mathematical structure is not a direct image of the physical form/process, but it represents the kind of information that is available for a perceiver and which is used for a recognition of the object. Consider three line segments that meet in a point P. P is not a structurally stable point for perception, in fact it has the highest degree of instability that is possible, because moving the gaze in any direction from P will lead to a qualitatively different configuration. But at the same time, P is the point that contains most information about the figure, because any small neighbor around P will contain a full replica of the whole figure. Catastrophe theory is therefore also a theory about the basic principles of perception. This has been confirmed by the research of D.D. Hoffman. It is not possible to go into details here, but for instance in Hoffman/Singh: Saliency of visual parts, they show that our perception prefers figures where the parts of the figure are separated by cusp points (the cusp being the most famous of the basic catastrophe forms).

Conclusion

Given the mathematical structures, there are two possibilities: either they are constructions of the human mind, which for some reason can be used to describe the physical world, or they reflect real properties of the physical world which the human mind is able to recognize. In for instance Lakoff/Nuñez: Where Mathematics Comes From, the authors argue for some version of the first possibility. René Thom on the other hand argues for the second. The mathematical forms in differential topology are precise expressions of the geometric and topological structures in the external world; this makes it possible for the biological creatures to recognize the external forms. It is for example decisive for perception that the forms contain singularities with compressed information, as illustrated above. In other words, there are structurally stable traits in the external world that correspond to structurally stable aspects of the perceptual process. The mind also consists of forms and they are in correspondence with the external forms. This is René Thom's extreme realism. It is of course easy to demonstrate aspects of human cognition that are not in a one-to-one correspondence with the external world. However, we have to understand René Thom right on this point. He is only speaking about processes in space-time that are accessible to our perception. The schematic forms of these processes, in terms of the critical points that define them, can be classified in differential topology, and it is because of these underlying schematics that it is possible for humans to recognize these processes. For this restricted area of human conceptualization, René Thom's realism might still be applicable.

References

Hoffman, Donald D. & Manish Singh (1997). "Saliency of Visual Parts", in: *Cognition*, 63 (1), p. 29–78.

Lakoff, George und Rafael E. Núñez (2000). *Where Mathematics Comes From: How the Embodied Mind Brings Mathematics into Being*. New York: Basic Books.

Thom, René (1972). *Stabilité structurelle et morphogenèse*. New York: Benjamin.

Thom, René (1980). *Modèles mathématiques de la morphogenèse*, Christian Bourgois, Paris.

Thom, René (1988). *Esquisse d'une sémiophysique*. Paris: Interéditions.

René Thom's Semiotics and Its Sources

PEER F. BUNDGAARD AND FREDERIK STJERNFELT

This paper introduces to the sources of René Thom's theory of meaning. It does so in an almost systematic, yet unhierarchical manner, by way of encyclopedic entries, as it were. However, the first four entries address the way in which Thom developed his catastrophe semiotics on grounds of his morphodynamical investigations in biology. This reflects, of course, our understanding of Thom's semiotics as being deeply rooted in his mathematical and biological thinking. We do evidently not claim to be exhaustive, as will be clear from our closing statements, but we do hope and aim to capture essential influences on René Thom's original, both genetical and structural determination of meaning, the nature of meaning, and the naturality of meaning – thereby also providing an understanding of the original synthesis which Thom's semiotics constructs from these sources.

Mathematics, Biology, and Meaning

As an integral part of his famous Theory of Catastrophes, René Thom developed, from the mid-60s and onwards, a semiotics. This semiotic theory develops out of the basic ideas of Catastrophe Theory (CT) to draw inspiration from a long series of theoretical sources in biology, philosophy, linguistics, semiotics, etc. This development has the main character that while the decisive impetus of Thom's semiotics comes from the mathematical-biological kernel of CT, the further development of the theory draws on an eclectic amount of inspiration from different sources. Main such sources are biologists such as C.H.Waddington, Jakob von Uexküll, Kurt Goldstein, Max Delbrück, Konrad Lorenz, d'Arcy Thompson, linguists such as Lucien Tesnière, Roman Jakobson, Kenneth.L. Pike, J. Ross, Hansjakob Seiler, philosophers such as Aristotle, Kant, Husserl, Charles S. Peirce, psychologists such as Wolfgang Köhler, and sociologists and anthropologists such as Frazer, Durkheim, Lévy-Bruhl and Gilbert Durand.[1] In this paper, we shall try to provide an overview over the most significant sources of René Thom's semiotics for two purposes: 1) to demonstrate and fine-tune his place in the recent developments of semiotics; 2) to argue for the originality and fertility of Thom's integrated semiotic speculations, disseminated in various papers and books as they remain. Thom's semiotics is intimately connected to his overall ideas concerning biology and epistemology. In a certain sense, it constitutes an original doctrine of biosemiotics, or, as Jean Petitot has it, a biolinguistics.

Thom once indicated the seminal idea of Catastrophe Theory[2]: he was at a conference at a German castle where he saw a series of large plaster castings depicting the different phases in the embryological development of a frog from the fertilized egg and to the grown-up, phenotypical creature. Thom immediately realized that this long process might be described as a continuous topological unfolding of discontinuities in a continuous medium, realizing the final, topologically complex shape which in some sense had been potentially present in the seemingly symmetrical simplicity of the fertilized egg. The egg, hence, was governed by a topological singularity, and the ensuing development could be described mathematically as the successive unfolding of that singularity. The beginning of CT, hence, was based on the idea that the topological structures Thom had studied in differential topology and singularity theory might be applied to the enigmas of phenotypical biological form which during the same period was slipping into the background in the focus on the genotype after Crick and Watson's DNA discovery. The mathematical core of CT is the so-called "splitting lemma" which states that in complicated differential equations in many variables, the singularities or the catastrophic points may be encapsulated in a small number of these variables. This lemma is essentially based on genericity arguments (it is improbable that many different variables stably reach non-

---

1 This indicates only main inspirations. Many more such influences can be traced, both negative and positive, such as biologists like Cuvier, Geoffroy Saint-Hilaire, Goethe, Lamarck; philosophers like Carnap, Cassirer, Derrida, Frege, Heidegger, Nietzsche, Leibniz, Russell, Stumpf, Wittgenstein; the linguists Benveniste, Chomsky, Granger, Greenberg, Harris, Sapir, Whorf; mathematicians as Brouwer, Riemann, Poincaré, Weyl, Zeeman, theorists of science like Boltzmann, Cavaillès, Cohen, Pierre Delattre, Hertz, Gerard Holton, Jean Largeault, Jean Petitot, Popper, Charles Renouvier; semioticians as Eco, Hjelmslev, Propp, Sebeok; psychologists and sociologists as Freud, Lacan, Lévi-Strauss, Weber. We shall intermittently touch upon some of them in what follows.

2 Personal communication.

generic catastrophic values at one and the same time). This gave Thom the basic idea in CT: viz. that in complicated processes in biology (and, later, in semiotics, sociology, anthropology, etc.), any abrupt changes taking place have their background in few variables only and may hence be charted independently of the complexity of the underlying process. This is why CT is essentially a phenomenological theory: it claims that in many interesting cases in biology, it is possible to bracket the detail of the underlying biochemical process and describe only the set of possible discontinuous changes ("catastrophes") on the surface level. The possible inventory of such sets of abrupt changes would then reduce to the seven "elementary catastrophes" – the simplest set of discontinuities in continuous, differentiable functions. These seven catastrophes are the only such sets of discontinuities which result from four or less variables alone. This argument, thus, initially depends on a basic spacetime realism, taking four simultaneous variables to correspond to one dimension each. This argument, of course, restricts the scope of the theory to the realization of the catastrophes in spatio-temporal form. Later, Thom was influenced by his British colleague Christopher Zeeman to modify and generalize this idea in a very important manner: to apply the CT formalism to the black box metaphor of system theory. If the underlying complex continuity was seen as a black box, the catastrophe variables need no longer be tied to spatiotemporal dimensions but might refer to any set of coexistent variables, and the resulting catastrophes hence taking place not in spacetime only (like optic caustics or embryology), but on all levels of more or less abstract behavioral spaces it is possible to chart in the scope of the theory, thus opening CT to modeling different behavioral discontinuities in biology, semiotics, sociology, political science etc. The base, however, remained: that CT permitted the distinction between the whole of a complex process, involving all variables and processing continuously in a slow dynamics, on the one hand – and the small, discontinuous part of the same process, involving few variables only and charting the quick, catastrophic changes of behavior. Zeeman's idea implied that the latter, catastrophic part of the system's behavior might even be described without any reference at all to the underlying dynamic and to the character of the variables involved. This gave CT the character of a phenomenological theory describing surface behavior of very different systems irrespective of their underlying physical or other substrate and immediately gave the theory a semiotic aspect: the possibility that stable aspects of the discontinuous surface phenomenology served as signs referring to the underlying more complex process.

C.H.Waddington

Thom's immediate application of catastrophe theory, however, concerned biology, and in a basic sense his semiotic core ideas spring out of catastrophe theoretical biology. One of Thom's basic inspiration was his friend C.H. Waddington and his concepts of "chreod", "canalization", and "epigenetic landscape".[3] Waddington coined this concept cluster in order to describe the epigenetic development of organisms which he did not see as an immediate effect of the genetic code. His idea was that the developmental possibilities inherent in the metabolism of the fertilized egg could be described by a landscape of potentialities where the development followed stable pathways of lowest energy – so-called "canalization" – what we would nowadays describe as the attractors of the dynamic system of metabolism. This model unites local determinism with global indeterminism on isolated spots in the landscape where the paths of lowest energy branch. (Figure 1)

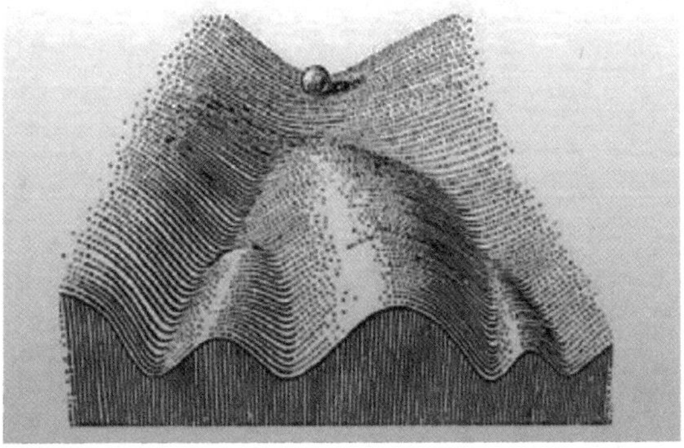

Figure 2. Waddington's epigenetic landscape. The path followed by the ball represents the development of an organ with its potential alternatives (branching points). As the ball rolls down it will meet a first alternative toward the left or the right, etc. Along each channel it will again be met by left/right alternatives at local bifurcation points.

Waddington's idea was that the function of genetic information was to influence the path taken at such bifurcation points – as a counterargument against the developing Neo-Darwinist doxa claiming that genetic information was in some sense the only and effective cause of the whole ontogenetic development. Waddington's model thus aimed at integrating metabolism (the general "landscape") and genetic information (determining the path at bifurca-

3 Waddingtonian ideas feature prominently in Thom's chef-d'œuvre *Stabilité Structurelle*, and during the writing of this work in the mid-60's Thom was in close contact with Waddington. Later, Thom summed up this inspiration in a short paper "An inventory of Waddingtonian concepts" (Thom 1987b).

tion points in that landscape), and his concept of "chreod" now referred to the tree structure of a distinct bouquet of such branching pathways in the landscape.[4] Thus, Waddington's landscape model offered a general geometrical, if speculative and non-specific, model which 1) gave an idea of how development integrated local determinism with (restricted) global indeterminism given by the set of branching possibilities, and 2) provided a model for the differentiation process taking place during epigenesis where the initial, undifferentiated fertilized egg in the course of successive generations of cell division differentiates into the around 300 highly different cell types in the mature organism; the branching points now being interpreted as simultaneous different pathways for different cell lines.[5] It is obvious how Thom saw the possible topological applications of his model. Waddington's potential landscape as such formed the catastrophic surface of the vast underlying complex continuity of metabolism, and the bifurcation points between pathways constituted the non-generic catastrophe points of the surface. Thus, Thom saw the possibility of CT providing parts of a Theoretical Biology, that is to say furnishing biology with qualitative mathematical models and morphological concepts ("chreod", "canalization", "landscape", "pathway", "branching point", etc.) transgressing the purely biochemical and genetic level. Especially the issue of cell distinction occupied Thom who had the idea that the spatial arrangement of cells and organs in the organism and their differentiation into subtypes formed two aspects of the same basic topological process. This immediately had semiotic implications because cell subtypes (different muscle cells, nerve cells, skin cells, bone cells, etc.) have immediate "semantic" interpretations tied to the purpose of the correspond-

4 In *Stabilité Structurelle*, Thom presented a generalized notion of chreod relevant for all dynamic processes, but later he claims that both Waddington's concept and this model were doomed from the beginning because of the impossibility of making them quantitative (Thom 1987b). This conclusion might seem strange, given the deliberately qualitative, non-quantitative character of much of CT in general. On other occasions, indeed, Thom seems rather unaffected by the non-quantitative character of the CT models. He often remarks that Zeeman was wrong in assigning such properties to them while remaining himself quite satisfied with their qualitative nature. The position is summed up in the title of his 1993 book *Prédire n'est pas expliquer*, that is: to predict (in terms of quantitative models) is not to explain (qualitatively).

5 As Thom also recognizes, there are thus two different interpretations of the bundle of pathways in the epigenetic landscape. One is temporal, the branching points indicating different parallel temporal development possibilities where the realization of one rules out the other(s). The other is spatial, the branching points now indicating a spatial splitting of the process into subprocesses, so as for instance different cell types.

ing organs. This is why Thom's semiotics begins already with ontogenetic biology. One of Thom's greatest achievements – according to his own judgment – is thus his semiotic interpretation of the earliest phases of epigenesis, leading from the sphere- and later ball-shaped morula over the cringing-in process to the bell-shaped blastula providing the first primitive inner-outer distinction in embryological development. From this, gastrulation takes place when the two germ "leaves" of the blastula are supplemented with a third, intermediary germ leaf. The innermost germ leaf of the gastrula now develops into the digestive tract, the intermediary leaf develops into the bone-and-muscle systems of the body, while the outer leaf develops into the brain and nervous system. During gastrulation, the intermediate leaf stretches out from the inner leaf and connects to the outer one – which Thom interprets as the hungry digestive tract which by means of the muscle and bone system catches a prey in the outer world (the prey being represented in the nervous system). Thus, the very intentionality of the organism is, according to this interpretation, already instantiated in the relation between the three parts of the gastrula in early epigenesis. Moreover, this form also instantiates the basic syntactic molecule in linguistics, the SVO formula, paving the way for Thom's idea that the understanding of a sentence is the unfolding of the simple form of the linguistic signifier into a tree structure, homologous to the unfolding of the fertilized egg in embryogenesis (and mirrorwise, the syntactic expression of a content as the "packing down" such a tree to germ form). In general, then, Thom has the idea that each bodily organ forms a stable threshold value in the unfolding topological shape of the body, so that each organ is from the beginning integrated as a specific phase in metabolism, regulating discontinuous catastrophic processes in the organism. This later becomes important for Thom's general interpretation of nouns in linguistics (see below). Metabolism thus becomes very important for Thom's outline of a theoretical biology: organs are seen as stabilizers facilitating catastrophic transformation processes in the organism without endangering its overall stability – the general stability of the organism in contrast to the ongoing chemical process of metabolism being the number one question in biology according to him. Thom's general idea of an animal, then, is that it is a complicated metabolism having a relatively small number of stable states. One such state is when all its basic needs are satisfied, but as soon as the animal due to ongoing metabolism lacks water, food, safety, sleep, sex, the state of the organism deteriorates to a lower stable state from which it has to perform a specific, catastrophic action to return to the optimal state. If it does not succeed, an even lower stable state awaits it in the limit: death.

Biologic metabolism is thus intertwined with the characteristic behaviors in which the animal must indulge in order to protect and continue metabolism – behaviors which have a characteristic discontinuous character (preying, mating, drinking, sleeping ...) and which may be described in simple catastrophes pertaining to the macroscopic state and behavior of the organism.

This is also why some of Thom's most important semiotic inspirations in his "biolinguistics" stem from early ethologists such as Lorenz, Pavlov, and von Uexküll.

Thom's theory of the sign

Thom's theory of the sign consequently differs much from the structuralist ideas of the arbitrary connection between signifier and signified, expression and content. To him, rather, signs are generally motivated: the signifier is motivated by its signified because of basic biological efficacity, and arbitrarity effects as found in language are derivative and later than the basic semiotic motivation. Signs emerge from animal behavior related to the large types of behavioral regulation (discovery of prey and predators, recognition of sexual partners, etc.). When the tick smells the butyric acid molecules announcing the presence of a mammal or when the predator recognizes the profile of its prey, the signifiers involved are engendered by their signifieds, Thom will say. The signifieds, however, are not only the mammal and the prey, respectively, because they are only rendered significant/they only carry meaning by virtue of the whole setting, by the whole functional circle involving them, as von Uexküll would have it (see below). It is rather the whole dynamic complex of the predator and the prey taken together which is the signified that engenders the signifier, in the same way as it is the whole epigenetic landscape of ontogenetic development which engenders the significant aspects which are the organism's organs. Thus, Thom's distinction between signified and signifier is built on his basic CT idea of the distinction between the underlying, slow, continuous dynamics which contains as an aspect the phenomenal, fast, discontinuous dynamics. The catastrophic surface is engendered by the underlying dynamics in exactly the same way as the signifier is engendered by the signified – in that way the whole of CT already forms a semiotics. It is the fact that the same basic processes give rise to the same typical catastrophic surface behavior which makes it possible for that surface behavior economically to function as a sign, making it possible for other complicated systems to relate to it in a simple, catastrophic way. This lies behind many of Thom's central dicta (say, "When you use the word information, you should rather use the word form" because it is the propagation of characteristic surface catastrophic form from one system to the other which constitutes information). This also lies behind Thom's basic distinction between nouns and verbs; the latter being basically characterized by referring to typical interactions between actants, while nouns refer to the actants as such. This is why Thom may say seemingly strange things such as that all nouns are basically animals (they are characterized by the different stable states they may change between, and by the kind of characteristic actions (describable by verbs) they may perform to so change between stable states). Thom tried formally to base this sign concept in a model where signs are seen as the propagation of simple, catastrophic forms between metabolic systems. This entails some interesting ideas: 1) the sign being necessarily energetically "small" in comparison to the involved systems, 2) the affected system necessarily possessing a receptor part with a high degree of sensitivity which may easily be coupled with the sign – as well as an "amplifier" able to pick up and reinforce such signs from the receptor system, 3) the necessity of a stable – non-chaotic – coupling between the two communicating metabolisms, necessitating the (relative) simplicity of the sign as well as its structurally stable character.

Jakob von Uexküll

The German-Estonian biologist Jakob von Uexküll appears early in Thom's semiotic essays. In fact as early as in the path-breaking "Topologie et signification" from 1968 where he is mentioned along with Kurt Goldstein as the main introducer of the concept of signification in biology prior to the DNA discovery. Their efforts "pour reconstituer les unités significatives du comportement animal ou humain" (in view of reconstituting the significant units of animal or human behavior) are, however, judged too finalistic. In Structural Stability and Morphogenesis (Thom 1972, English translation 1989), this judgment seems changed, and especially Uexküll's late Theory of Signification is referred to. Uexküll is quoted for the idea that

> The mechanism of any machine – consider for example a watch – is always built up centripetally, meaning that the various parts, the cogs, springs, and fingers, must first be completed and then mounted on a common support. As opposed to this, the growth of an animal, for example of a Triton, is always organized centrifugally from its germ which first develops into a gastrula, and then constantly sprouts organ buds. There is an underlying construction plan in both of these cases; for the watch, the plan controls a centripetal process, in the Triton, a centrifugal process. It seems that the parts are assembled according to diametrically opposite principles. (Thom 1989, 254)

The idea is that animals and machines share the same, basically teleological nature, but have quite different construction principles; the former being centripetal in the sense that all parts must be produced before assembly, while the latter are centrifugal because epigenesis takes place by the unfolding from a germ in successive distinctions creating the single organs so that parts are continuously differentiated out of a pre-existing whole.

Only later, Thom widens the reference to Uexküll's theory of signification, most often in connection to his own development of the doctrine of "saliency" and "pregnancy" (see below). Saliencies are perceived physical Gestalts while pregnancies are biological significations which an organism, for innate or conditioned reasons, may ascribe to such salient forms. Pavlovian conditioning, now, is described as the possibility of associating a new saliency (the bell sound) with a pregnancy-bearing saliency (feeding), thus investing this pregnancy in another form, causing appropriate action to be performed (salivating) when the new saliency only is presented. (Thom 1981). The original pregnant form (the presence of edible objects) is seen as significant in the sense of von Uexküll's theory of certain objects possessing biological signification for an organism (cf. Thom 1986) – because the perception of such forms triggers large scale hormonal and motor reactions in the organism. Pregnancies may thus spread in chains of associations from one saliency to the next – by the Frazerian laws of similarity and proximity – so that pregnancy may spread "in the phenomenal field of experienced forms as an invading fluid which propagates through the those cracks in the field which are constituted by the salient forms."[6]

Uexküll's theory is his famous "functional circle" which places the organism in the flux of two types of signs, the incoming "Merkzeichen" (perception signs), the outgoing "Wirkzeichen" (action signs), related to the same object which the organism only keeps contact to through these signs.

6 "[...] dans le champ phénoménal des formes vécues, comme un fluide invasif qui se propage à travers les fissures du champ que constituent les formes saillantes." (Thom 1991a, 68).

Thom does not explain this theory in detail, but it is evidently the idea that the organism is able to respond adequately to a few select environment objects to secure its own stability only if it is able to interpret the relevant perception signs and respond with the relevant action signs. Uexküll's tick description is – just like in Uexküll himself – taken as the prototypical example, and Thom repeats it on several occasions. Uexküll's import on theoretical biology is identified with the "pregnancy" in Thom's developed version of his theory (see below). The tick waits, according to Uexküll, on a leaf or a twig for the smell of butyric acid which immediately makes it let go and fall in order to hit (hopefully) the fur of the mammal for which the acid served as a sign. With a rudimentary sense of heat, it crawls to the hottest spot on the skin of the animal (where the skin is thinnest) in order to be able to bite through the skin, suck blood and procreate by parthenogenesis. As Thom writes, the reproductive cycle of the tick is thus tied to a very small number of signals (or indices, Thom 1991a – rather, in fact, they will be symbols in Peirce's terminology) which Thom calls "prégnances" in his biological use of the word as opposed to the mere physical Gestalts of saliencies.[7] Only certain olfactory and thermic forms are relevant signs in the tick's *Umwelt*. Thus, the tick example serves to show the degree to which meaning is rooted in early biological development, long before man: "Any stability, any homeostasis in the living being rest on the possibility of almost instantaneous recognizing which are of vital importance and of responding adequately to them."[8]

7 Without mentioning it, Thom's distinction between pregnancy and saliency goes into an internal tension in Uexküll's work. In his earlier writings, so as for instance his *Theoretische Biologie* (1920/28), Uexküll saw that higher animals have large capacities for perceiving "Merkzeichen" *without* any immediate biological signification (as for instance, in their charting and mapping of their environments) while only certain, privileged objects within that environment are able to release the catastrophic activity in the organism and thus qualify as part of the "functional circle" of the animal. In the later *Theory of Signification* (Uexküll 1940), the simple function circle of the tick is taken as prototypical and generalizable for all organisms so that all animal perception is taken to be significant – thereby deleting the earlier distinction between significant and non-significant forms. Thom's development, interestingly, goes in the other direction: while taking the older Uexküll's simplistic theory as his point of departure, Thom, as it were, reinvents the more sophisticated distinction of the younger Uexküll, saliency corresponding to the "neutral" objects of higher animals, while pregnancy is that biological signification which only certain of such objects possess.

8 "Toute stabilité, toute homéostasie chez l'Être vivant reposent sur la possibilité de reconnaître quasi-instantanément les stimuli d'importance vitale, et d'y répondre pertinemment." Thom 1991b, p. 2.

The combination of Uexküll's theory of meaning with the Gestaltist notions of pregnancy and saliency, consequently, gives a biological basis to Thom's theory of meaning, founding, in turn, both his linguistic and his epistemological ideas.

Structural Linguistics (Ferdinand de Saussure, Roman Jakobson)

Elmar Holenstein (1992) once reported that Roman Jakobson had considered the three greatest structuralists to be Prince Trubetzkoy, Claude Lévi-Strauss, and René Thom. Thom, in contrast, does not seem to have been sufficiently influenced by Jakobson so as to express any corresponding acknowledgement or gratitude. The same goes for his relation to Saussure (or to any other major structuralist for that matter, with few exceptions, cf. below). In the Oeuvres complètes (Thom 2003), Saussure and Jakobson are each mentioned in about 20 texts, and mainly, almost exclusively, in general terms (of the sort "the arbitrarity of the sign" according to Saussure, or "the paradigmatic and syntagmatic axes of language" according to Jakobson, etc.). In fact, when going through his first texts on language, one cannot help getting the impression that first Thom developed his own theory of language (his topological syntax), only upon which he discovered linguistics.

On a couple of occasions however, Thom makes quite extensive use of both Saussure (negatively) and Jakobson (positively) to develop two fundamental aspects of his theory of language, which simultaneously make out two fundamental epistemological problems in linguistics in general: (1) the problem of developing a "morphology of meaning" or (of "the signified" as he puts it with reference to Saussure) in view of establishing an intelligible relation between language as a symbolic form of expression and the contents expressed by language (a "de-arbitrarization" of the sign system, as it were); (2) the problem of overcoming the antinomy between description and explanation, structural and genetic, formal-immanent and causal-transcendent determinations of language. Let us briefly take a look at both of the two issues.

In "Topologie et linguistique" (chapter 11 in Thom 1980a), Thom remarks that if we adopt a Saussurean view on language and thus define it as a system of signs, then only one aspect of the sign seems amenable to objective description, namely the physical one, the signifier, while the signified, being accessible only through introspection, escapes all third-person determination. In Thom's eyes, this state of affairs is paradox, if not simply an aporia, insofar as our folk understanding of language and the function which language serves calls for a robust link between what we have in mind (the signified, or our meaning intentions, as Husserl would have put it) and the expressive form that articulates it (the signifier). In short, it seems crucial to determine the passage: signified → signifier. Now in a Saussurean framework and by virtue of its herostratically famous "arbitrarity of the sign" this issue can simply not be addressed. As Thom concludes, in the last analysis, this dogma entails that everything about language can be described and explained except how and why it articulates meaning.

The problem raised here is probably one concerning the *encoding* of pre-linguistic representations in language. It has two aspects: 1) representations, or the signifiers, are endowed with a form, a structure which precedes their linguistic articulation; 2) if such representations are to be expressed faithfully, language must dispose of means (semantic forms) in virtue of which it can express these structures. From such a perspective, the Saussurean transformation from amorphous signified → formed signifier, which in this framework remains opaque (to say the least), can be readdressed in the following terms: we can consider the level of the signified as a structured domain, a *morphology of meaning*, and thus consider the signified-signifier relation as a *mapping* of this morphology onto the morphology of language. The issue is then no longer addressed in terms of the arbitrary link between signifier and signified, but in terms of a re-articulation of one order in another order. In other words: the fact that the relation between signifier and signified is arbitrary at a local, intra-sign level, does not imply that the relation between the sign system and its object is globally arbitrary, since the former is supposed to simulate the latter. Inspired by Zeeman, Thom has the idea that pre-linguistic content may be seen as a manifold in many dimensions R^n (in turn possibly referring to a state-of-affairs with similar or even greater dimensionality). If such a content is to be linguistically expressed, it must be cut up into R^4 pieces able to be rendered in single minor events. These events, in turn, must be projected onto the linear one-dimensional string of syntax so that formally, the encoding of content is a mapping $R^n \rightarrow R^4 \rightarrow R$, necessarily losing some of the content originally present, in order to attain the simplicity and stability of syntax. This analysis of semantic content into pieces accessible to syntactic representation of course requires a mirror process of synthesis to take place in the receiver.

Now, the semantic forms are the linguistic counterpart to the experienced structure. As regards the crucial representation of events, these are, Thom claims, evoked by the verbs whose schematic meaning simulates the external event type they designate or refer to (cf. below on Lucien Tesnière). Obviously, the validity of Thom's solution to the problem rests on the possibility of producing a consistent determination and delimitation of the semantic forms and their corresponding event types. On the other hand, the way he states the case seems to address a key issue in any research program within semiotics: 1) Man is characterized by his capacity to organize sense data into pre-linguistic representations of the external world which are faithful simulations of core aspects of the spatiotemporal processes they refer to. 2) Whatever the origin of language is, one of its fundamental functions must have been to communicate relevant information about things and events in the environment. 3) Thus, language must be capable of faithfully describing basic structures of spatiotemporal events and must therefore dispose of means to encode such information (a morphology of meaning). 4) It is one of the tasks of linguistics to explain how the structure of our representations is encoded in language, in terms of a morphology of meaning.

In short, Saussure is the, as it were, negative backdrop of Thom's realistic conception of the relation between language and thought, and between thought and the structural makeup of the objects it refers to. The Saussurean lexicon allows him to stress the difference between the arbitrarity inherent to a morphology of the signifier (in its relation to the signifier), and the motivated nature of the morphology of the signified (in its relation to structure in the environment).

Now, as regards Roman Jakobson, the only time he is invoked in non-general terms and in a rather extensive quotation (something Thom is far from being accustomed to) is in an epistemological discussion which is at the heart of Thom's early ventures into the domain of linguistics. The discussion concerns the classical *Genesis*/*Geltung* problem (origin vs. validity) and thus addresses the classical antinomy between casual-reductionist and formal-structural approaches to language; an antinomy which Thom, in the vein of Jakobson, sets out to overcome. Roughly speaking, the antinomy – or the aporia – consists in the sorry alternative between considering language

a) as a non-autonomous system and explaining it in terms of some other system of constraining properties (relative to, say, the organism which uses language, its psychology, its (neuro)-physiology; its phonatory and auditory system, its ecology, history and sociology, and so on): in which case the object – Language – is denied any ideal, general properties (it is no object proper, no *Gegenständlichkeit* as Husserl would have put it), and it is as a result reduced to being an epiphenomenon of the positive existence of its material cause;

b) as a self-contained formal symbolic system determinable solely in terms of its own intrinsic relational or structural properties, independently of any extrinsic reality, purpose, function, or relation to other cognitive domains (perception, reasoning): in which case you may obtain a relatively faithful description of language – on the grounds of a purely immanent grammatical a priori – which has a negative counterpart, though: the incapacity of motivating and explaining the emergence and the very nature of the formal structures governing the linguistic system: "where do the structures come from?" as Thom enjoyed asking.

It is while evoking the Scylla of reductionism and the Charybdis of autonomism that Thom (1974) refers to Jakobson's cognate concern:

> […] if we do not thoroughly combine the complementary notions of autonomy and integration, our attempt will be led astray: either the prosperous idea of autonomy will degenerate, as all particularisms, into an isolationistic and self-destructive prejudice, i.e. a sort of separatism or apartheid, or we go in the opposite direction, whereby we compromise the sound principle of integration by substituting the indispensable autonomy with an infelicitous heteronomy (or 'colonialism') (Jakobson, *Essais de linguistique générale*, II, p. 256; our translation).

Catastrophe Theory as applied to linguistics is in Thom's view a way of overcoming this state of affairs. On the one hand it supplies with a morphological description, thus doing justice to the phenomenological autonomy of language as a higher order object (again: a Husserlian *Gegenständlichkeit*), on the other hand it is as such a theory that aims at motivating the nature of the morphological structure of language, since the structure is determined in terms of the externally realized and syntactically simulated conflict between two or more forces/actants. As Thom himself claims:

> For the catastrophe theoretician the structure is no "*a priori*" fact, it does not belong to some Platonic empyrean. It is a direct result of the conflict between two (or more) forces which generate and sustain it. This makes it possible to develop a classification of forms as well as an algebra, a combinatorial system of forms on a multidimensional space […] We are now close to perceiving the outline of a dynamic structuralism which reintegrates

> causality and time and thus appears as a general theory of forms which is independent of the specific nature of the substrate space. (Thom 1974, p. 245, our translation)[9]

Lucien Tesnière (and Aristotle)

Thom endorsed a basic form of Aristotelian realism throughout his whole intellectual life. This is, of course, explicitly the case in his Sketch of a Semio-Physics (Thom 1988) and its outline of a common sense ontology based on a reinterpretation of Aristotle's Physics as a charting of naive physics, but it is already key to his early conception of the relation between mind and matter. In his view, perception is a question of extracting form or structure, i.e. morphological information, from the matter in which it is realized. Perception is, roughly speaking, a "transfer of form" from matter to mind, and this is indeed to be taken in a literal, strong sense: the state into which the mind stabilizes, corresponding to the understanding of the experienced object, is a state which rests on the same structure as the object itself; to perceive is to take in a form from one domain, the physical realm, a form which is then re-unfolded in another realm or domain, the soul or the mind. Hence, Thom's Aristotelian punch line: "the message which has an autonomous meaning inherits its structure from the external catastrophe it intends to mean" (Thom 1972, p. 329); the content of the intentional act is thus claimed to possess the same structure as the external object it refers to. This entails that the meaningful representational states in the brain are governed by the same morphogenetic principles as the phenomenal manifestation of their reference objects (things and processes). If this is so – that is if such Aristo-

9 Pour le théoricien des "catastrophes" la structure n'est pas donnée "a priori", elle ne sort pas d'une empyrée platonicienne. Elle est directement issue du conflit entre deux (ou plusieurs) forces qui l'engendrent et la maintiennent par leur conflit même. Ceci permet de développer une classification des formes, ainsi qu'une algèbre, une combinatoire des formes sur un espace multidimensionnel; ainsi, les possibilités de l'approche structurale se trouvent-elles considérablement augmentées, et en expliquant la morphologie par un dynamisme sous-jacent, on peut rompre l'antinomie des tendances réductionniste et structurale. On peut, en effet, faire une théorie rendant compte des mécanismes de cause, et classifier les formes archétypes qui passent par couplage causal d'un substrat à un autre. Ainsi, s'entrevoit la possibilité de créer un structuralisme dynamique, qui, réintégrant la causalité et le temps, se présenterait comme une théorie générale des formes indépendante de la nature spécifique de l'espace substrat.

telian hylemorphism obtains, then, obviously, a major task consists in laying down the types of structure (or catastrophes, to use Thom's term) which organize the interaction between spatial entities and thus make out the scaffolding of those general processes and events humans perceive and conceptualize. In its semiotic versant Catastrophe Theory sets out explicitly to establish this: it aims to lay down the principles which govern the semiotic interface between phenomenal forms and representational forms.

Now, this formal isomorphism between the structure of objects (in the broad sense, including event types) and the structure of perception transposes directly to Thom's topological theory of language. The structural form of the object is not only re-unfolded as content of perception, but also in and by the structure of the symbolic system in which such contents are expressed, i.e. language. In other words, Thom claims that the principle organizing the combination of meaning-carrying units in language corresponds to the principle underpinning the configuration of phenomenal parts into intelligible wholes in perception. The rationale of this claim is biological: it seems sensible to suggest, as Thom says (1980a: 180), that language has evolved from the necessity of (or the advantage inherent in) conveying to others the significant changes (i.e. the catastrophes) in the environment. This entails – as Thom with no further argument asserts – that the syntactic structure "naturally" reflects the dynamic structure of the external catastrophe. But how, then, is the dynamic structure of the external catastrophe actually encoded in language? To which extent can it make sense to claim that one and the same principle of organization should be active in two realms as apparently different as the linguistic and the perceptual?

Here, Thom is standing on the shoulders of the founding father of modern syntax, Lucien Tesnière (1893–1954), who in 1934 published the article "Comment construire une syntaxe" which preceded his monumental work on structural syntax, posthumously published in 1959 with the title *Eléments d'une syntaxe structurale*. As a matter of fact, Tesnière counts among the most important linguistic influences on Thom, as he puts it rather bluntly: "Personally, only two linguists have taught me something: Lucien Tesnière (Bloomfield's contemporary as regards his use of graphs) and Hansjakob Seiler, German linguist who founded a universalist group in Köln."[10]

10 "Personnellement, seuls deux linguistes m'ont appris quelque chose: Lucien Tesnière (contemporain de Bloomfield pour son emploi des graphes) et Hansjakob Seiler, linguiste allemand qui a fondé un groupe universaliste à Cologne" (Letter to Monsieur Lekkari Abdelbasset, March 26th 1991; Inédits 1991, p. 105, in Thom 2003).

In the following, we shall very briefly present some of the tenets of Tesnière's syntax, establish the affinities to Thom's topological syntax as well as point to a couple of differences.

Tesnière defines syntax as a system of dependence relations with superordinate and subordinate lexical entities. Syntactical analysis, thus, consists in laying down the hierarchy of connections between lexical entities in a sentence. In 'my old friend', 'friend' is the superordinate word, whereas 'my old' are the subordinate terms ('old' being the superordinate of 'my'). In 'my old friend Alfred reads a book', the highest superordinate term is 'reads', whereas 'Alfred' and 'book' are the immediate subordinate words with each their subordinate expressions. Now, key to Tesnière's theory of syntax is the notion of *valency*. As just suggested, Tesnière places the verb or the verbal "knot" on top of the hierarchy of connections. It is the core element of linguistic expressions, and Tesnière defines it in terms that are readily transposable to recent grammars' use of argument-role structure and the like. "The verbal knot," Tesnière says, "expresses a real small drama. As a drama it indeed comprises a process, and most often actors and circumstances." (Tesnière, 1959: 102, our translation). Now, the number of "actors" – or to use Tesnière's own and in continental semiotics widespread term: "actants" (semantic roles) – attached to a verb defines its "valency" or argument structure. Two elements connect directly to Thom's topological linguistics as developed in e.g. (Thom, 1980a, originally 1971).

a) The structural principle of combination underlying both Tesnière's and Thom's theory of syntax is not purely formal-linguistic. Linguistic entities are not combined by virtue of their categorical form (analytically attributed with some valency), rather they are combined correlatively to the schematic content expressed by the verb: it is, thus, not a formal property of the verb "give" that it has a trivalent structure (giver → gift → givee), it is a (real formal) property of the scene it refers to. Part of Tesnière's legacy in Thom could indeed be defined in terms of the intrinsic semantic import of syntax. Here, syntax is less a principle for the linearized combination of lexical entities by virtue of their categorical form than it is a principle for the configuration of lexical entities by virtue of their semantic structure. Seen from the point of view of contemporary cognitive linguistics (Talmy 2000, Langacker 1987–1991), both Tesnière and Thom more or less explicitly anticipate the crucial intermediary schematic level, shared by both language and perception: accordingly, verbs evoke a schematic representation of a "small drama" or express as their genuine content the "phenomenological catastrophe" in the scene they designate: syntax simulates the interaction between spatiotemporal actants. This also tends to motivate the strong constraints syntactic structure is submitted to, as we suggested above: if the highest value of verb valency is 4, this is, according to Thom, due to the fact that the interactions they represent are themselves constrained by the maximum number of simultaneous minima in the seven elementary catastrophes. Thom's theory thus provides a sort of minimalist cognitivist idea; however, this aspect of his theory – the outline of a schematic ontology roughly defined in terms of the 7 elementary catastrophes and the archetypal graphs – is exactly what has lead to Jean Petitot's reframing and re-elaboration of morphodynamic semiotics in view of hooking up the "deep" morpho-topological structures that Thom laid bare with the "shallower", but descriptively much more efficient conceptual schemata developed in cognitive science.

b) More technically, Thom's linguistic theory could be characterized as a topological schematization of the concept of valency. Thom construes Tesnière's connection between actants in terms of positional connections between places occupied by actants in an abstract space (to this, cf. Petitot 1985, 1992, 1995). The structural connections underpinning surface syntax which Tesnière defined in terms of his "stemma" are thus redescribed by Thom as positional configurations: a trivalent verb consequently encodes a connexion between three positional actants. Such positional configurations are limited in number since there is a limited number of possible undecomposable interactions between actants. Thom's well-known archetypal graphs or interactional graphs are models of the fundamentally possible forms of interaction between actants. His aim is therefore to objectively determine the nature and number of event types expressed in and by language. If, according to Thom, the message inherits the structure (the catastrophe) of the state of affairs it refers to, then, at a global level, this structure exerts a strong constraint on language: indeed, the nature and limits of event types necessarily defines the nature and limits of linguistic combination of meaning units. The set of event types (catastrophes) serves as a global characterization of the sentence structure in general. At a local level, in any given expression, a verb can be said to unfold a schematic landscape with actants interacting in a general way (the *interaction type* of the verb) and comports a specification of the interaction (the *action type* of the verb, as it were): the ar-

> chetypal graph captures, of course, the general schematic make-up, say of a *transmission* type, whereas the particular verbs provide the concrete flesh: 'give', 'hand', 'send', 'offer', 'donate', etc. The *interaction type* provides the global structural unity of the expression; the *action type* its local specificity.

It is easy to evaluate the extension of Tesnière's influence on Thom, but it is difficult to assess its depth. On the one hand, Tesnière is the most recurrent reference in Thom's semio-linguistic writings: he appears for the first time in the seminal "Topologie et signification" (1968, reprinted in Thom 1980a), never to disappear again (Thom still warmly refers to Tesnière and the influence he exerted on him in his 1997-talk "Graphes-Signaux-Prégnances"). Moreover, there are, as we have just suggested, important analogies between Tesnière and Thom. Firstly, of course, both rest on a mereological-semantic definition of language structure in terms of dependency relations between superordinate and subordinate parts (with the verb as the kernel and other word classes as satellites hooked onto it) – i.e. they both marshal they idea that in language the structural configurational order prevails over the linear, combinatorial order.[11] Secondly, and perhaps less known, Thom has acknowledged his "diagrammatical" depth to Tesnière to the extent that the Strasbourg linguist was the first to propose "*a graph-based structural* explanation of grammatical structure," that is to say a schematization of sentence structure that does not simply mirror or reproduce its linear order.[12]

On the other hand, however, Thom's own theory of syntax is quite rudimentary with respect to Tesnière's detailed descriptions and fine-grained distinctions. His references to Tesnière are in almost all cases general in scope and basically boil down to simply evoking Tesnière's concept of valency.

---

11 Tesnière's syntactical structure has an evident semantic ground insofar as its argument" or "actant" structure is relative to the type of event it refers to; it is therefore not a simple analytical property of a given categorical form: the ditransitivity of the verb "give" is something proper to the kind of event it refers to, not an *a priori* property of that verb as a categorical form.

12 Cf.: "Des théories linguistiques que j'eus l'occasion de rencontrer, celle de Tesnières m'avait particulièrement plu. Car ce fut la première à expliquer la structure grammaticale par une *structure en graphe* de cette partie du discours qu'on appelle usuellement la *phrase*" (Among the linguistic theories I had the occasion to encounter, I was particularly fond of Tesnière's. It was indeed the first to explain the grammatical structure of the part of discourse we usually call the *sentence* by a graph structure (Thom 1997, p. 31).

It seems thus sensible to contend that Thom's theory of language is kindred, intellectually affine with Tesnière's, but not built on his grounds. In fact, Thom's assumptions as regards language are, both generally (for example, as to the relation of language to perception and the formal structure of the phenomenal world) and internally (as to its own formal make-up) a direct consequence of his ontological commitment. In other words, with or without Tesnière, Thom would most probably have come up with the same sort of theory. Once you claim that the central issue to be addressed within linguistics is how language can syntactically simulate the structure of the events it refers to, and once you claim that these event structures cluster in determinable types, then the rest follows: there must exist recurrent semantic forms which can re-articulate this interactional structure, and there must exist lexical entities in charge of expressing this sort of content, viz. verbs. In contrast to Tesnière – who was busy doing other things – Thom explicitly aims at bringing linguistic structure back to the world.

Saliency and pregnancy (Wolfgang Köhler)

The development of Catastrophe Theory is known to constitute the first part of Thom's philosophy of nature. The concept of pregnancy is, along with its counterpart "saliency", the pillar of what he himself called his "second philosophy", that is to say the one devoted to the characterization of the "intelligible ontologies" and relaxing the dependency on the mathematical formalisms of the original Catastrophe Theory (Thom 1988 is the main reference as regards this endeavor). Three things should be mentioned in this respect: (1) Thom introduces the notion of pregnancy, albeit not saliency, in the heyday of Catastrophe Theory, cf. for example the article "De l'icône au symbole" from 1973 (Thom 1980; Eng. transl. 1985). We are consequently dealing with a constant in Thom's work, perfectly compatible with the basic catastrophe-theoretical framework.[13] (2) However, in "De l'icône au symbole"

13 This is indeed worthwhile mentioning since sorry attempts have been made – perhaps even to some extent by Thom himself – to pull down an iron curtain between the "structural" catastrophe-theoretical period (with its Heraclitean emphasis on the stabilizing *logoi* sustaining the order of the phenomenal world and its representations) and the post-catastrophistic period with its Parmenidian emphasis on the continuum and the slow flow

(1980a, originally from 1973), Thom distinguishes between biological pregnancies (i.e. the capacity of a form to evoke other biologically significant forms) and physical pregnancies (i.e., the intrinsic, salient, structural stability of a form). The notion of biological pregnancy clearly anticipates the conception of pregnancy Thom will marshal in the 1980s, so that biological and physical pregnancy simply correspond to the later notions of pregnancy and saliency, respectively. (3) Whereas Thom's own use of the concept is perfectly clear, the relation between Thom's understanding of "pregnancy" and its Gestalt-theoretical origins in Max Wertheimer and Wolfgang Köhler (Thom's main, if not simply only reference in this respect) remain rather unclear.

The text "Prédication et Grammaire Universelle" (Thom, 1980b, but written and presented in public in January 1978) seems to be one of the very first to introduce the notion in its by now classical conception: "Let's say that a spatial form (*F*) is 'pregnant' if its perception causes great physiological or behavioral response in the observer (human or animal!)" (Thom 1980b: 3). However the first text to be systematically devoted to this topic is probably "Morphologie du sémiotique" (from 1981, reedited in Thom 1992). Here Thom defines saliency as a distinct, detached phenomenon; a visual form is an epitome of this, but an acoustic phenomenon would do all as well, a tactile too. A salient form may take on a specific signification for a biological organism, i.e. provoke a hormonal reaction and an emotional response. This is, as already mentioned, what Thom understands by the pregnancy of a saliency. So the prey of a hungry predator is highly "pregnant" for the predator. Essentially, this sense of "pregnancy" is how French and English dictionaries define the expression (as "fullness" or "richness" in a lot of different respects). It means something very different in its German version, in a sense it even means the inverse here, namely "concise", "precise", "clear". Now, this would be anecdotic if, on two occasions, Thom himself had not referred to Wolfgang Köhler and Gestalt Theory as the origin of the concept such as he uses it. First in the mentioned article from 1973: "The fact that a theory of the pregnancy of forms is possible was the essential doctrine of Gestalt The-

---

of pregnantial meaningfulness. As an argument against this specious divide one could point to Thom's all as constant emphasis on the interdependency or fundamental correlation of discontinuity and continuity: meaningfulness implies necessarily morphological detachment,or saliency, which then in turn is necessarily a detachment from and thus an articulation of a continuum. So even though, at least from the 1980s and on, Thom assigns ontological primacy to the continuum, obviously he still holds that any theory of form, any morphology in whatever domain, must rest on the irreducible distinction between regular and irregular points, and hence on the concept of discontinuity.

ory that W. Köhler marshaled with courage and lucidity" (Thom 1980 (1973): 264). And bis repetita many years later where Thom in a footnote remarks that his use of the term is "obviously" an upshot of the German "Prägnanz", due to Gestalt Theory in general and Max Wertheimer and Wolfgang Köhler in particular (Thom 1988, 32 note 4).

The latter quotation of course echoes the former, yet none of them justifies Thom's claim to the effect that he uses the concept of pregnancy in Wertheimer's or Köhler's sense. Wertheimer's laws of Prägnanz (Wertheimer 1923) are indeed principles ruling the organization of forms on purely perceptual grounds, independently of all biological interests and categorical meaning. These principles, which govern the "grouping" of qualities according to proximity, color, closure, size, orientation, "good continuation," etc. do, of course, have a biological rationale: they further fast object recognition, but they do not concern an animal's chemical and emotional reaction to what it perceives – its tendency or not to consider a given form as biologically significant – and for this reason they do not sustain pregnancy in Thom's sense. Rather, Prägnanz in Gestalt Theory strictly applies to saliency as Thom defines it: they are the laws that explain how qualities are grouped and organized so as to form detached figural wholes in perception. Notice also that the prime property of pregnancy according to Thom is its unbounded, continuous "flow" character, whereas the prime character of gestaltic Prägnanz is its closed, discontinuous character. The "flow" is, as it were, what makes it possible for pregnancies to invest Prägnanzen; the latter's discontinuous, bounded character is what makes it possible to contain the former. This is by the way probably the reason why Thom, in his 1987-paper on the epistemology of evolutionary processes, explicitly notes that he uses the term "pregnancy" in the French sense of "prégnance" (Thom 1987a, note 1).

The concept couple plays a central role in Thom's idea of "ontologie intelligible", his formal requirement for any regional ontology, as it were. The idea is that any ontological domain must possess concepts of these two types; simple, stable, spatially well-defined forms (saliencies) and meaning-bearing forces able to propagate between saliencies and to inhabit and modify these saliencies in stable and characteristic ways (pregnancies) – so to speak forms and forces, to put it crudely. The very notion of "intelligible ontology" is, of course, in some sense a *catachresis* uniting epistemology and ontology in one expression. Thom's basic realism, however, seems to preclude a pure epistemological reading of it, pointing rather to his idea that all ontologies *must* be understandable because of their necessary inclusion of these two elements. To that extent, Thom's use of the word couple, while originating in Gestalt psychology, takes it far beyond that in order to form the basis of an audacious general principle of formal ontology. The intertwinement between epistemology and semiotics indicated in Thom's famous motto "What must be sought is a theory of language which immediately involves a theory of knowledge" ensures that this "intelligible ontology" is tightly connected to his semiotics: the duplicity of saliency and pregnancy in ontology is mirrored by the duplicity of noun and verb in linguistics.

Kenneth L. Pike (emic/etic)

Kenneth L. Pike plays an episodic role in Thom's theory of language. Roughly speaking, Pike simply provides him with a distinction (emic vs. etic) which Thom, in turn, adapts to his own framework in order to capture a property of language that he considers as a grammatical universal, namely the distinction between more speaker independent or "emic" entities in contradistinction to the "etic" features of language which are more speaker dependent. The emic-etic terms constitute each their pole on one of the axes which organize and structure the universal categories (word classes) of grammar. At one end, the *emic* end, we thus have the most speaker-independent terms: the epitome of which are Nouns. Thom considers them as quasi-organisms possessing their own regulation system, with external limits protecting their signification against "aggressions" from cognate terms pertaining to the same semantic domain or network of systematically correlated terms. The signification of a concept has thus a wide temporal scope, it is unlikely to undergo changes, and it is, in its default understanding, quite autonomous with respect to the pragmatic situation of communication. At the other, *etic* end we have the fundamentally speaker-dependent categories, such as deictics, whose meaning is unilaterally dependent on speaker's and hearer's spatiotemporal anchoring. A couple of variations notwithstanding, Thom parses the whole continuum from emic to etic in the following way: nouns, verbs, adjectives, possessive pronouns, numerals, and deictics (Thom 1978a, 1980b; Pike 1971).

Algirdas Julien Greimas

The founding father of European post-Hjelmslevian semiotics, A.J. Greimas, is only sparsely mentioned in Thom's work (The CD-Rom containing the complete works of Thom notifies 14 references, of which many are purely incidental). He seems to have had no import on Thom's semiotics – quite contrary to what was the case for Thom's disciple Jean Petitot – and the only aspect of his work which Thom refers to (with no bibliographical specifications) is the "semiotic square" (the first attested reference to it appears in an essay entitled "Modèles mathématiques de la morphogenèse" from 1971, partially included in the 1974-version of the book with the same name and left out of the 1980-version). Here Greimas' semiotic square is qualified as one of those "very gross objects" which the structural methods within the human sciences use to do their formalizations.

Thom mainly deals with the square in its narratological version, i.e. as a formalization of an ordered, cyclic sequence of transformations of states or narrative turning points (naturally interpreted as "catastrophes"). His reinterpretation of it consists 1) in stripping the semiotic square of its "logical clothing" ("Structure cycliques en sémiotiques", Thom 1992, p. 73), by 2) proposing a dynamic modeling of its formal, logical scaffolding in terms of the cyclic hysteresis structure described below.[14] Though there does not seem to exist any deep theoretical affinities between Thom and Greimas, the project of developing dynamical interpretations of the formal models, already sketched in the above mentioned article from 1971, has of course a major historical signification: it anticipates Jean Petitot's general schematism and naturalization of structure, which initially was fulfilled in terms of a topologization of the semiotic square (Petitot 1982, 1985, 1992).

14 This example of topological schematization of logical structure is a constant in Thom's thinking: no matter the domain (mathematics, philosophy or linguistics), the aim "is not to provide Geometry with a logical foundation, but to found logic [*le* logique] on Geometry" (Thom 1988, p. 16).

Joseph Greenberg

It seems natural that Joseph Greenberg's search for linguistic universals should appeal to a semiotic theory with the universalist and realist ambitions like Thom's.

Especially Greenberg's basic syntactic typologies (SVO, SOV, etc.) prompt Thom to launch a detailed theory of sentence construction and understanding ("Sur la typologie des langues naturelles" (from 1970, reprinted in Thom1980a)). As mentioned above, the initial idea was that a complex signified be decomposed in (at most) four-dimensional units. How are these units now further syntactically analyzed? Thom's idea is that the relevant manifold is subjected to analysis by the continuous lowering of one parameter, counting the attractor minima as they are reached, thereby generating a tree structure (somewhat like the possible tree structure in a chreod section of a Waddington landscape) – this tree structure now forming the relevant syntactical tree for expressing the content in question. Following Tesnière, the most basic part of such trees will be the verbal part, privileging the VOS structure (rarely realized in reality) from the sender's point of view -because the information of the verb kernel of an utterance gives access to the scene connecting the actants and circumstants. The reason this syntax is not realized, Thom argues, is the receiver's point of view for which the opposite order SOV is the most natural, with VSO and especially SVO as compromise possibilities. The receiver's first interest is taken to be S – who is responsible for the action presented in the sentence, an important issue in the prototypical case where the sentence relates to states-of-affairs in the immediate space-time surroundings of the interlocutors. This also prompts Thom to his analysis of the genitive: in expressions like "Le chien de Jean", X's Y, Jean (X) is "semantically destroyed" because the whole of the expression refers to a dog, and only Jean's ownership is left of him in the expression. The genitive thus makes room for any possible relation between its two relatees and only indicates their coexistence in some space – and thus facilitates abstract thought as liberated from any particular actant graph – : one of Thom's ideas as to the human semiotic privilege over animals which are supposedly always caught in the actual actant graph of their present behavioral purpose.

Peirce

References to Peirce's work occur stably in the later Thom. Among Peirce's detailed philosophy, they refer to two themes especially: Peirce's doctrine of categories and the most well-known taxonomy of his semiotics, that of icon-index-symbol. The initial reference to Peirce takes place in the 1972 paper "De l'icone au symbole" (in Thom 1980a) which outlines a whole theory of the man-animal difference on the basis of an interpretation of that "second trichotomy" between icons, indices, and symbols. This triad of concepts distinguishes between the three ways in which the object of the sign may be indicated, namely either by means of similarity between sign and object, of an actual connection between them, or of a habit or convention. Thom's basic observation is that all three of these are present already in animals, taking the roots of rationality beyond any simple idea of a sharp borderline between man and animal. Thom argues for the importance of iconic signs – as against a tendency to see them as trivial – because even simple images contain a tension between reversibility and irreversibility. Images are irreversible because some information is lost in the mapping, and the same image may refer to several objects, and, conversely, the signified engenders an indefinite number of such signifiers of itself. On the other hand, a restricted reversibility is possible because, in some cases at least, it is possible stably to recover the signified from the signifier. The information thus recoverable is that which is structurally stable and hence able to survive the noise inherent in any transaction – the physically pregnant features (the later saliencies). Indices have the biological importance of leading, spatio-temporally, to the object (important in the case of biologically significant objects, of course) and indices for that reason have the ability to form chains (if one index leads to another which leads to the object), founding associationism. Symbols, finally, are tied to the threshold points between two attractors in the life of the animal where a decision must be made whether to pursue the regulation leading into the other attractor or not: here attractive and repulsive symbols lead in each their set of attractors. How can the purpose of the sign engender its morphology? – This is, to Thom, the basic question of semiotics as such (or, as he here says, of semiology). The answer is "the principle of the backwards path" which takes 1) the spatio-temporal localization of the sign as first prerequisite: the sign informs about the localization of the object, 2) aspects of the virtual result of the encounter of the object (the cranium as a sign of danger, knife and fork as a sign of a restaurant coming up) which is projected "backwards" toward the sign interpreter. The multiple possibility for such aspects – and in general the multiplicity of different indices referring to any catastrophe, such as the different pieces scattered on the road after a traffic accident – accounts, in Thom's view, for the Saussurean arbitrarity of the signifier. Thom's interpretation of the icon-index-symbol triad is developed further in Thom 1980c, where the vector leading from signifier to signified is considered in relation to time. The three Peircean sign types are taken, not unrelated to Peirce's theory, to refer to the signifieds in the past (index), present (icon), and future (symbol)[15], respectively, which gives Thom the occasion to argue against theories privileging the signifier over the signified or even reducing away the latter (deconstructionism, Hilbertianism). Symbols are taken to arise out of the chain-forming capabilities of indices along with a temporal reversal (Pavlovian conditioning again taken as prototypical example), conforming, as it were, to Peirce's idea of the symbol's "esse in futuro". Both papers terminate with speculations about the difference between animal and human semiotics, claiming that the latter is characterized by the weakening of the source forms of biological pregnancies so that disinterested representations and the long, branching chains of human sign use arise. This part of Peirce's influence thus serves to detail the salience-pregnancy hypothesis.

The other Peircean influence thus concerns Peirce's category table – Firstness-Secondness-Thirdness – which catches Thom's interest in connection with cognitive and linguistic universals. Thus, Thom has an idea, foreign to Peirce, that there is a general syntax of thought and expression following these categories:

> Firstness, Secondness, Thirdness. Exemple français: ça sent le brûlé.
>
> La triade peircéenne Qualisign – Sinsign – Legisign comparée à
>
> Prégnance – Saillance – Concept (Nom).
>
> (Firstness, Secondness, Thirdness. French example: It smells like burnt. The Peircean triad Qualisign – Sinsign – Legisign compared to Pregnancy – Salience – Concept (Noun) ("Reflexions sur le continu de Hansjakob Seiler", 1992, 2, in Thom 2003)

The idea is that "Ça" (it) represents undifferentiated Firstness, "sent" (smells) indicates the Secondness catastrophe of perception confronting subject and

15 Points also developed in "Morphologie du semiotique" (1981, in Thom 1992), and "Mathématique et sémiotique" (Thom 1981).

object and “le brûlé” (burnt, however nominalized in French) specifies the kind of Thirdness regularity encountered in the perception – while Peirce’s First Trichotomy is taken to reflect the process of making a stable concept type out of what was first only a vague pregnancy.

These ideas, however interesting, are not further developed, and occur as clarification attempts in the context of Thom’s discussion of the second linguist, besides Tesnière, who influenced him the most, the German universalist linguist Hansjakob Seiler with whom Thom collaborated during many years.

Hansjakob Seiler

One of Seiler’s main ideas is that of linguistic continua orienting and governing a long range of linguistic phenomena. Seiler has an example of the continuum from Deixis to Predication which Thom very often cites, viz. the long German chain of predicates: “Diese erwähnten zehn schönen roten hölzernen Kugeln” (“These above-mentioned ten beautiful red wooden balls”).[16] The idea is that in such an adjective series, other constraints on syntax vanish, so that an underlying continuum going from subject towards object becomes visible; the adjectives spontaneously placing themselves on such a continuum going from the most subjective: deixis, and to the most objective: the noun, over adjectives referring to discourse, number, subjective and objective qualitites. Thom’s reinterpretation of this continuum goes as follows:

> It seems to me that the fundamental findings of Professor Seiler, about the continuum of determination could be stated as follows: any place in the standard syntactic tree which gives rise to a subordinate predication may be exploded into an auxiliary continuum (a “technique”); this technique then involves an arbitrary number of terms, but the ordering of these terms is not arbitrary, it is governed by order associated to the semantic nature of the corresponding invading “pregnancy”.
>
> The order seems to be the following:
>
> - Subjective generated pregnancies, originated by the speaker as: deixis.
> - Mental pregnancies associated to subjective appreciation of the speaker
> - (subjective adjectives)

16 Cf. W. Wildgen 1999.

- Physical qualities of accidental nature (e.g. color) (objective adjectives)
- Material nature
- Genitive – spatio-temporal localization (Thom 1983c, 254).

Thom's claim is that this continuum of Seiler's no less than solves Kant's classic issue of schematism by continuously connecting subject and object. Thom's central claim is that this subject-object continuum is identical to the emic-etic axis (see above) but independent of another dimension which he sometimes attempts to describe as the continuum of semantic complexity, necessitating a two-dimensional representation of universal grammar which he gives several different outlines of ("La double dimension de la grammaire universelle", 1978a; another takes the degree of freedom of pregnancy propagation to substitute for semantic complexity, "Contribution", 1983; in Thom 2003) without necessarily assuming that two dimensions may be sufficient. The general idea, thus, is that universal grammar requires a number of independent continua for its description. The existence of such overarching continua makes Thom state his anti-formalist credo as to syntax: "The generativity of generative grammar (Chomsky's competence) is probably no more than a formalistic illusion." (Contribution 1983, p.12; in Thom 2003). Seiler's theory thus permits Thom to embed his syntactical deliberations as motivated by overarching structures of semantic continua and leads him to many further interesting hypotheses, partwise in collaboration with Seiler.

The anthropology of hunting – Gilbert Durand

In the semantic end of Thom's semiotics lie his surprising reflections on animal hunt as involved in the origins of symbolism. The main ideas rest on an Uexküll-like idea of the circular structure connecting animal metabolism with outer action, locating signs in the animal's search for prey. The animal must be capable of recognizing certain signs for its prey (predators, sexual partners, rivals, etc.), and thus also be able to imagine and fantasize about these Gestalts:

> However, there can hardly be any doubt about the fact that any animal "fantasizes" its preys (and its predators), for otherwise it would be unable of recognizing them. The imagined "Gestalt" most certainly precedes and shapes the real Gestalt (as it is by the

way shown in animal etiology by the phenomenon known as "supranormal" releasers). (Thom 2003: Letter to Gilbert Durand Jan. 27 1976, Inédits 1976; in Thom 2003)[17]

Gestalts which have a certain idealized quality documented by Lorenz' famous discovery of "supranormal releasers" like the artificial version of the seagull mother's beak with two red spots (real seagulls having only one), able to arouse the young more than any empirical beak. During the three consecutive versions of the chapter 10 of Stabilité structurelle (the French 1972 version, the English version, and the French 1977 version), Thom refines this basic idea, and in "Les racines biologiques du symbolique" (Thom 1978b), integrating a model based on a cyclic path around the organizing center of a Fold catastrophe on the one hand with, on the other, the French philosopher and anthropologist Gilbert Durand's doctrine of symbolism in his Les structures anthropologiques de l'imaginaire (1960). The former part follows the below model:

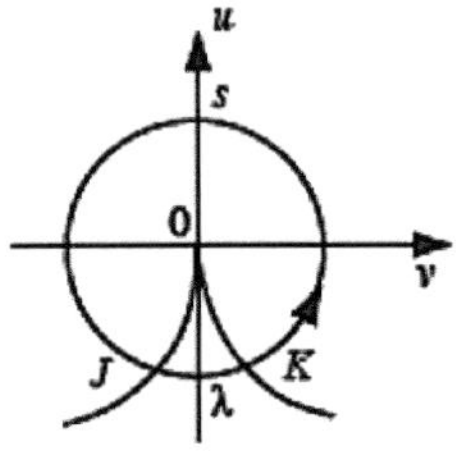

Figure 3.

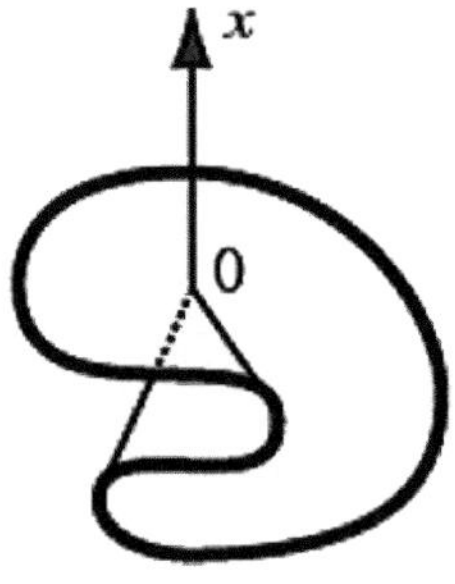

Figure 4.

Figure 2. Thom's hysteresis model.

17 "Il ne fait guère de doute, cependant, que tout animal "fantasme" ses proies (et ses prédateurs), sans quoi il ne pourrait les reconnaître. Très certainement, la "Gestalt" imaginée précède et "informe" la Gestalt réelle (comme le montre d'ailleurs, en éthologie animale, le phénomène bien connu des déclencheurs "supranormaux")."

These two figures show the circular process changing between the presence of one and two actants. The upper figure shows the basic catastrophe "cusp" (the pointed figure consisting of two discontinuity lines meeting at the origo). The circle indicates the movement in the cusp landscape, and points J and K indicate the two catastrophe points where the circular trajectory intersects with the cusp. Inside the cusp, that is between J and K, two minima exist which in Thom's interpretation corresponds to two actants. Outside the cusp, only one minimum exists. These minima, in Thom's interpretation of the hunt, now correspond to the predator and the prey. The lower figure shows the cusp as the projection of a fold onto the u,v plane of the upper figure of the fold. The bold line indicates the circular trajectory in the fold landscape – and the two catastrophes occur where the trajectory shifts direction in the folding-in of the circle. In the upper figure, the phases of the hunt are as follows: in its hungry state, the animal is between the v-axis and the point J on the trajectory. Only one minimum exists here, corresponding to the predator, alienated in the image of its prey. J constitutes the perception catastrophe where the image acquires outer existence in a prey and the hunt begins. During the hunt, in the phase between points J and K, two minima exist, corresponding to both predator and prey. The chase culminates in the capture of the prey in K, after which alienation ceases. Now in the post-K phase, the predator consumes its prey and thus becomes one with it, only one minimum again exist. In the upper part of the tracjectory, the predator sleeps and digests its prey and so wakes up again, hungry and again alienated, and the process starts over again. Thom's interpretation of Durand now finds the roots of daytime symbolism in the J-K section inside the cusp with its clear, binary predator-prey distinction, evident valorisation and themes like friend-foe, war, victory, defeat, etc. The other, upper, part of the cycle, then corresponds to nighttime symbolism with blurring of binary categories, identification of predator and prey, dreamlike descent into another worlds, etc.

It is based on the idea that there is a tension between two different identity notions in biology – the persistence of the animal subject as opposed to its repeated, different actions. The idea of such a tension rests on the unspoken presupposition that animal minds are tied to the present now and hence must change according to the behavior type they indulge in. Therefore, the spatiotemporal identity of the animal is opposed to the different mental-behavioral "identity" states of the animal. In the model, this conflict of identities is construed according to the following interpretation of the path indicated: in the phase of the path approaching the catastrophic part, the animal is alienated by the image of the prey, it is dominated by the urge to find a real counterpart to match the image. The first catastrophe indicates the actual perception of prey fitting the image, and the short interval to the next catastrophe is the hunt where the animal has regained its proper self. The second catastrophe indicates the second discontinuity: the catching of the prey; doing so, the animal finally reaches the normal state and becomes its satiated self. During sleep, in the smooth half of the pathway, the animal gradually again, by means of digestion, internalizes the prey: it becomes its own prey, as Thom liked to put it. And when it wakes up hungry, alienation is active again. Durand's doctrine distinguishes between by a diurnal and a nocturnal symbolism in human imagination which Thom identifies with the two separated phases of the cycle. The hunt corresponds to Durand's daytime symbolism, covering rise, fight, and victory – or being hunted, if the identification goes with the prey – in general, this symbolism is articulated in clear binarities. The digestion phase now corresponds to Durand's nighttime symbolism, covering the *coincidentia oppositorum* where the two sides of binarisms flow into each other, an imagery of descent into unclear worlds of subjective dissolution dominate. Finally, the structure taken as a whole may give rise to cyclic imagery, exchange and dialogue ideas. As is evident, the whole model is at the same time taken, once again, to mirror embryogenetic structures, the clear distinction between predator and prey being analogous to endo- and ectoderm.

The biological semiotics integrating Peirce, Gestalt Theory, and embryology is here extrapolated to constitute a basis for general anthropological imagery involving the destruction and resurrection of the self.

Different small influences

In addition to the main inspirations which we have charted here, a series of smaller ones may be listed. A negative one is Jacques Derrida, highlighted by the fact that Thom was among the originators of the 1992 protest address against his nomination as a honorary doctorate at the University of Cambridge (along with Quine, Ruth Marcus, Barry Smith, David Armstrong, and others), and in Thom's works the idea that one sign, allegedly, refers to the next in a never-ending sequence is taken as the opposite to his own realist semiotics: "a theory that is very popular in the Parisian salons claims: 'there is no signified, only signifiers; each sign refers to other sign in an endless regression'" ("De l'icône au symbole", Thom 1980a).

Edmund Husserl is sometimes referred to as a subjectivist who, along with Heidegger, has destroyed French epistemology, at other occasions he is reconsidered as anticipating Thom's theory of the basic status of hunt and caption (*Erfahrung und Urteil* is interpreted as saying that "recognizing the object is nothing but grasping it" ("Temporal Evolution of Catastrophes", Thom 1973, p. 42)), or he rightfully defends Euclidean geometry as the necessary basis of all other geometries, or he is taken to support the idea that the recognition of a shape constitutes consciousness rather than the opposite (also with reference to *Erfahrung und Urteil*, cf. "Préhension et perception", 1992, p. 165). Apart from being used as a subjectivist scoundrel, Heidegger is mostly credited with the one-liner "Die Wissenschaft denkt nicht" and its subsequent "and it is not its purpose either." which Thom quotes with negative acclamation, complaining about this divorce between (rigorous) science and (meaningful) philosophy.

Bernhard Riemann is several times quoted for the idea: "[...] Riemann, who 'committed' a few philosophical pages, in which you encounter the following essential idea: if a mental process, taken in its neurophysiological reality, is described by a set, a variable variety W(t) of the space of neurophysiological states of the brain, then the (subjective) signification of that process is defined by the form (in a sense which is still to be made clear) of the variety W in the space of the states" (Thom 1983, p. 195)[18] – a very catastrophe-theoretical idea indeed; that the signification of a thought is constituted by the form of the underlying neurophysiological process. A related idea is Paul Valéry's idea from *Eupalinos* which Thom quotes over and over; here, Socrates is portrayed as saying that a form is geometrical if it can be finitely described in few words; which basically means that relative geometrical simplicity and the ability to serve as symbol are related.

In the epistemological end of his semiotic musings, Thom often refers to the theorist of science Gerard Holton and his doctrine of "themata": i.e., his ideas that scientific discoveries are often prompted by deep epistemological (and unsolvable) thematic tensions such as unity-diversity, order-disorder, permanence-change, simple-complex, holism-reductionism, for-

18 "Riemann, qui 'commit' un petit nombre de pages philosophiques, dans lesquelles on peut trouver cette idée essentielle: si un processus mental, dans sa réalité neurophysiologique, est décrit par un ensemble, une variété variable W(t) de l'espace des états neurophysiologiques du cerveau, alors la signification (subjective) de ce processus est définie par la forme (en un sens à préciser) de la variété W dans l'espace des états."

malism-empiricism, conventionalism-realism, continu-disconitnu, etc. This basic semantic inventory for scientific (and other) imagination leads Thom to his doctrine of "*apories fondatrices*", the idea that different scientific disciplines have, as their organizing center, an unsolvable antinomy (tied to its partition in saliencies and pregnancies, forms and forces) constantly urging scientific discovery onwards (geometrical continuity vs. algebraic discontinuity in mathematics, empty space vs. physical entities in physics, metabolism vs. organismic stability in biology, the semiotic inability for sign systems to talk about their own foundation in linguistics etc.). The overall conclusion being:

At any rate, the only way we can attain an explanation, a generation of empirical diversity from one unique principle is by resorting to a process of emanation, of "procession", as the Neo-platonicians saw very well. If this process is assimilated with a propagation of pregnancy (accompanied by a convenient oscillation between saliency and pregnancy), we will be led to look for the "source forms" of this pregnancy: a black hole which can only be filled by a fantasmatic image, a local and temporary solution of the founding aporia) (Thom 1992, p. 481).[19]

Explanation must take the shape of the propagation of pregnancies across saliencies, and the relevant science is called to the impossible task of finding the unitary origin of that pregnancy. Thom later discovers (from Jean Largeault) that largely similar ideas had been proposed by the French Neo-Kantian Charles Renouvier already in his *Les Dilemmes de la Métaphysique Pure* from 1901.

Concluding remarks

Our intention has been to introduce to the major sources of Thom's semiotics and the way they fit into his original biosemiotics or biolinguistics. Thom's

19 "Dans tous ces cas, le seul espoir d'arriver à une explication, un engendrement du divers à partir d'un principe unique, est de faire appel à un processus d'émanation, de procession comme l'avaient bien vu les néo-platoniciens. Si ce processus est assimilé à la propagation d'une prégnance (assortie d'une oscillation convenable entre saillance et prégnance), on sera ramené à chercher les "formes-sources" de cette prégnance: un trou noir qu'on ne pourra jamais remplir que par une image fantasmatique, solution locale et temporaire de l'aporie fondatrice."

semiotics develops from its topological and biological core to span linguistics and epistemology, semantics and ontology, and during unfolding it grafts upon this germ a growing number of influences, adding to both the richness, the complexity and the inner tensions of Catastrophe Theory semiotics.

References

Durand, G. (1960). *Les structures anthropologiques de l'imaginaire*, Dunod, Paris.

Holenstein, E. (1992). (interview with Roberto Benatti) Phenomenological structualism and cognitive semiotics, *Scripta Semiotica* (1).

Holton, Gerald (1973). *Thematic Origins of Scientific Thought: Kepler to Einstein*, Harvard University Press, Cambridge, MA.

Jakobson, R. (1969/1976). *Essais de linguistique générale*, t. I–II, Minuit, Paris.

Peirce, C.S. (1931–58). *Collected Papers*, Thoemmes Press, London.

Petitot, J. (1985). *Morphogenèse du sens*, PUF, Paris.

Petitot, J. (1992). *Physique du sens*, Ed. du CNRS, Paris.

Petitot, J. (1995). Approche morphodynamique de l'iconicité des stemmas. In F. Madray-Lesigne and J. Richard-Zapella (eds.). *Lucien Tesnière aujourd'hui*, Editions Peeters, Louvain: 105–122.

Pike, Kenneth L. (1971). *Language in a unified theory of the structure of human behavior*, 2nd rev. ed., Mouton, The Hague [u.a.].

Renouvier, C. (1901). *Les Dilemmes de la Métaphysique Pure*, Alcan, Paris.

Saussure, F. de (1915). *Cours de linguisitique générale*, Payot, Paris.

Seiler, H. (1995). Diversité des langues et conceptualisation: le cas de la détermination nominale, *Intellectica*, 1/20: 127–137.

Stjernfelt, F. (1992). *Formens betydning. Katastrofeteori og semiotik* ("The Meaning of Form. Catastrophe Theory and Semiotics"), Akademisk, Copenhagen.

Stjernfelt, F. (2007). *Diagrammatology. An Investigation on the Borderlines of Phenomenology, Ontology, and Semiotics*, Springer, Dordrecht.

Tesnière, L. (1959). *Eléments de syntaxe structurale*, Klinksieck, Paris.

Thom, R. (1972). *Stabilité structurelle et morphogenèse.* Paris: Ediscience.

Thom, R. (1973). L'évolution temporelle des catastrophes. Collection *Applications of global analysis, I. Symposium Utrecht State Univ.*, Utrecht, 7–9 février 1973: 61–69. Comm. Math. Inst. Rijksuniv. Utrecht, 3, 1974.

Thom, R. (1974). La linguistique, discipline morphologique exemplaire, *Critique* (322), 235–245.

Thom, R. (1978a). La double dimension de la grammaire universelle. *Circé*: 78–90.

Thom, R. (1978b). Les racines biologiques du symbolique. *Circé*: 40–51.

Thom, R. (1980a). *Modèles mathématiques de la morphogenèse*, Christian Bourgois, Paris.

Thom, R. (1980b). Prédication et Grammaire Universelle. *Fundamentae Scientiae* 1/23: 23–34.

Thom, R. (1980c). L'espace et les signes, *Semiotica*, 29, 3–4: 193–208.

Thom, R. (1981). "Mathématique et sémiotique", in Thom 2003.

Thom, R. (1983a). *Paraboles et catastrophes*, Flammarion, Paris.
Thom, R. (1983b). La modélisation des processus mentaux. In *Fonctions de l'esprit,* Hermann, Paris: 192–206.
Thom, R. (1983c). Transitivity Continua and Prototypicality. In H. Seiler & G. Brettschneider eds. *Language Invariants and Mental Operations*, Günter Narr Verlag, Tübingen:256–263.
Thom, R. (1986). Les chemins du sens à travers les sciences. *Sciences et Symboles*, Colloque de Tsukuba, 6–10 novembre 1984, M. Cazenave, éd., Albin Michel, Paris: 361–380.
Thom, R. (1987a). Epistemology of evolutionary processes. In W. Callebaut & R. Pinxten (eds.). *Evolutionary Epistemology*, D. Reidel Publishing Company, Dordrecht: 97–105).
Thom, R. (1987b). An Inventory of Waddingtonian Concepts. Conference at the Waddington Memorial Conference, Oaxtepec, 1–5 septembre 1987, Mexique. *Theorical Biology. Epigenetic and Evolutionary Order from Complex Systems*, B. Goodwin & P. Saunders eds., Edinburgh, Edinburgh University Press: 1–7. (In Thom 2003).
Thom, R. (1988). *Esquisse d'une sémiophysique*, InterEditions, Paris.
Thom, R. (1989). *Structural Stability and Morphogenesis*, Redwood City, California etc.: Addison – Wesley Publishing Company, inc.
Thom, R. (1991a). Saillance et Prégnance. *L'inconscient et la Science*, R. Dorey éd., Dunod, Paris: 64–82.
Thom, R. (1991b). Relativité du Vrai, Relativisme de l'Intelligible, *Epistemologia* 15 (1): 3–20.
Thom, R. (1992). *Apologie du logos*, Hachette, Paris.
Thom, R. (1993). *Prédire n'est pas expliquer*, Flammarion, Paris.
Thom, R. (1997). Groupes – Signaux – Prégnances. (Document nb. 4.2, "Inédits 1997", in Thom 2003.
Thom, R. (2003). *Oeuvres complètes*. CD–ROM edited by Institut des Hautes Etudes Scientifiques. Bures-sur-Yvette, France.
Uexküll, J. v. (1920/1928). *Theoretische Biologie*, Springer, Berlin.
Uexküll, J. v. (1940). Bedeutungslehre (= Bios, Abhandlungen zur theoretischen Biologie und ihrer Geschichte sowie zur Philosophie der organischen Naturwissenschaften. Bd. 10), Verlag von J. A. Barth, Leipzig.
Waddington, C.H. (1957). *The Strategy of the Genes*, George Allen & Unwin, London.
Wertheimer, M. (1923). Untersuchungen zur Lehre von der Gestalt II, in *Psycologische Forschung*, 4: pp. 301–350. Translation published in Ellis, W. (ed.) 1938. *A Source Book of Gestalt Psychology*, Routledge & Kegan Paul, London: 71–88.
Wildgen, W. (1999). *De la grammaire au discours. Une approche morphodynamique*, Peter Lang, Bern-Berlin-Wien.

Thom's Theory of "Saillance"and "Prégnance" and Modern Evolutionary Linguistics

WOLFGANG WILDGEN

1. Introduction

René Thom, the famous mathematician and critic of current science was only attracted by semiotics and linguistics indirectly. At a point in his career as a mathematician,[1] being appointed to the Institute of Advanced Research in the Sciences (Institut des Hautes Études Scientifiques) in Paris, he began to reflect on the role of topology and topological dynamics in the fields of physics, chemistry and biology. In the 60s he exchanged letters with the English theoretician of biology C.H. Waddington, who wrote a preface to the first (French) edition of Thom's book "Stabilité Structurelle et Morphogenèse" (1972) and to its translation into English (1975).

In the second preface Waddington refers to his own book published in 1940 "Organizers and Genes" (cf. Thom, 1977: XVIII), where he formulated some fundamental questions of theoretical biology, to which Thom found a mathematical answer. One problem of biology is that it cannot be explicitly founded in laws of physics or chemistry. Such a foundation would ask for systems with millions or billions of factors. A topological treatment allows us to jump (theoretically) over these complexities and to formulate the general contours of an explanation (with all the risks such a jump implies). In his foreword to the English translation of Thom's book (1975), Waddington says that the last chapter of Thom's book, "From animal to man: thought and language", is the most stimulating. The reader of Thom must be surprised to see topics from physics linked to problems of human communication and language (ibidem: XIII). The picture Waddington gives is clear-cut: Thom is a mathematician (topologist) normally concerned with questions of theoretical physics, who applies his structural intuitions to the fields of biology, semiotics and linguistics and for this purpose neglects the restrictions which

1 Thom got the Fields medal for his mathematical work in 1958.

govern scientific discourse in these disciplines. In a much later book, written after the catastrophe controversy (around 1978) and the shift of the theoretical avant-garde to chaos theory, Thom wrote his book on semiophysics "Esquisse d'une sémiophysique" with the subtitle "Physique aristotélicienne et théorie des catastrophes" (1988). His modern Aristotelism which is opposed to the Galileian world view and his renewal of a "Naturphilosophie" in the spirit of Goethe and Schelling were the philosophical horizon, into which he could embed his theoretical thinking. His theory of "prégnance", which is my topic in this contribution, pertains to this period of his research (i.e., after 1978), and one of my basic concerns will be to explain why and how it stands in a continuity with the models proposed in his book published in 1972.

2. From differential topology to „semiophysics": Thom's intellectual itinerary

Thom saw language in a quasi continuity with biological development (morphogenesis). As in the case of biological morphogenesis, the specific mechanisms (genetic, chemical or neural) are not yet known (since then some of them have been discovered, however). Thus the model proposed by Waddington and elaborated by Thom was rather global and concerned the major topological features and the process of differentiation as such (bifurcations, appearance of barriers, separation of specific organs and subfunctions, etc.). In this realm, Thom could apply the new results obtained in catastrophe theory based on the classification theorem proven by Mather in 1964. The broad view of these concerns is summarized in his book (ready in 1968, published in 1972) "Stabilité structurelle et morphogenèse". At this point the geometry of the umbilics was not yet known in detail, and in the second (French) edition Thom added some remarks concerning the umbilics, but he did not modify the list of semantic archetypes, although those derived from the umbilics could not be maintained under the new insights. In my habilitation thesis of 1979 (cf. Wildgen, 1979 and, for the published major chapters, Wildgen, 1985), I gave a more correct derivation of the sending-archetype from the compactified elliptic umbilic and showed that the archetypes with more than three minima have to be derived from the (non-elementary) double cusp (X_9) and that other possible types have to be considered. The classification of paths which had been done intuitively by Thom is also problematic for all catastrophes beyond the cusp (cf. Guckenheimer and Holmes, 1983). Meanwhile, the catastrophe controversy, which regarded mainly the applications proposed by Christopher Zeeman (cf. Zeeman, 1977, for a collection of his articles and Sussmann and Zahler, 1978, for the controversy), had weakened the international interest in applications. Therefore, the necessary elaboration of catastrophe theoretic semantics (cf. Wildgen, 1982, Petitot, 1985, and Wildgen, 1985) was not thoroughly reviewed and referred to in the community of applied mathematics. The semiotic and linguistic hypotheses announced by Thom were only taken up by some linguists (e.g., Jakobson, Seiler), but not systematically pursued.

In the same period (after 1978), Thom went one step further and tried to specify the forces which govern the process of semiosis and not only the topology of its outcomes. In the frame of his training as a mathematician primarily concerned with applications in physics, he first tried to link the forces of the morphogenesis of meaning (semiogenesis) to known basic forces like gravitation, radiation (light), etc. These universal fields embed the living beings and govern their environment (ecology). They are naturally the background of all perceptual and motoric processes. In perception, light is at the basis of our visual perception; gravitation underlies human and animal motor-processes and the sensation of pressure and weight; sound waves are registered by the ear and the diffusion of chemical substances evokes reactions of our taste and smell organs.

As the dynamics of such fields (e.g., light) have been the topic of physics since Newton, and wave dynamics were the topic of specific mathematical treatments since Maxwell, it was straightforward for Thom to postulate a specific field that is registered and filtered by our sensory organs. He called it "saillance", i.e., those effects which stand out may be selected as informative in a psychophysical field. His program was to extract as much systematic content as he could from the analogy between physical fields and perceptual fields. Insofar as perception is the basic stratum of semiosis, any perceptually based symbolic structure elaborates the psychophysical fields of our sensory systems.

Although this strategy: from physics to semiotics allowed for the transfer of many mathematical techniques, which had been shown to be successful in physics, there remained a large gap between psychophysics (the level of perception) and linguistics (or cultural semiotics). The term "prégnance" had to fill this gap and to explain the transition between very basic perceptual reactions and language (culture).

This very coarse summary of René Thom's itinerary between 1965 (exchange of ideas with Waddington) and the eighties (elaboration of the theory of "prégnance" and other cultural applications) has to be understood as the background for my more specific comments and proposals. I shall begin with the sources Thom could use for the elaboration of his concepts: "saillance" and "prégnance" (cf. also the contribution by Bundgaard and Stjernfelt in this volume).

3. The concept of "prégnance" (vital relevance)

The basic meaning of the term introduced by Thom comes from German "Prägnanz", which has been internationalized by Gestalt-psychology. In German the adjective "prägnant" means that something perceived, enounced, remembered has outstanding properties, which catch attention, make it relevant, important in a specific situation; one also associates brevity (laconism) and sudden effect as in jokes or aphorisms with it. Other words, that may be associated to this lexical field, are "prägen", "Prägung", i.e., the coining of a piece of money or the imprinting of a sign on an object. "Prägnanz" also has a *positive* value in Gestalt-psychology insofar as the term "gute Gestalt" (optimal form) refers to "Prägnanz". The term in French "prégnance" (or even less evident in English "pregnancy") should be understood as a lean-translation with the lexical content of German "Prägnanz".[2] In the following I shall use the English term "pregnancy" with the meaning of "prégnance" given by Thom.

The dichotomy introduced by Thom has a precursor in Aristotelian philosophy, and it is Peirce who pointed to this grounding in 1905, although he did not use the terms "pregnancy" and "salience"; he rather distinguishes the principle of growth (or germ; the female principle) versus the principle of impact (or chock; the male principle). The Aristotelian notions for this duality are: dynamis (δύναμις) or energeia (ἐνέργεια) for the female principle, and morphe (μορφή) and typos (τύπος) for the male principle. If we come

2 In the etymology of "prägnant" and "Prägnanz" one may distinguish two different sources: a) before "prae" + be born "(g)nasci" in Latin, with the figurative meaning: heavy, full; b) the etymology of "prägen" (imprint), medieval German: praech (en). The modern meaning blends both sources.

back to the original meaning of pregnancy, we are very close to the notion Peirce takes from Aristotle, i.e., the female principle, the germ and the principle of growth. The female principle is furthermore associated by Peirce to the mind (the available consciousness), whereas the male principle is associated with experience, it is the confrontation with reality (typos is translated as the hitting on something which resists, i.e. reality). Thom was not aware that he was thinking in the footsteps of Peirce, but both thinkers had Aristotle as the firm ground of their theoretical thinking; thus the coincidence does not astonish.[3]

Thom refers explicitly to a number of modern scientific currents which contain a similar dichotomy. In the scientific use of the two terms, concepts stemming from different traditions may be distinguished:

a) In Pavlov's experiments, dogs salivate if presented with meat and "learn" the conditioned reflex of an associated bell (which evokes similar reactions). Here the concept is used implicitly; the meat and the bell are "pregnant". The meat is "pregnant" by virtue of the biological constitution of the dog, its body shows a set of automatic reactions which are thus the content of the presented sign "meat". The associated sign "bell ringing" inherits the meaning of the sign "meat", and classical stimulus-response theory postulates that all meanings are derived by stimulus conditioning from basic reflexes.[4]

b) Konrad Lorenz observed the process called "Prägung" (imprinting) in birds (e.g. geese). During a short period after they left the egg, many birds select rather unspecific stimuli in their environment and quickly elaborate them to form basic concepts like that of a "mother bird". Thus,

3 Peirce also mentions autobiographical sympathies for a Schelling-like conception and thus he shares a grounding in the Goethe-Schelling tradition also considered by Thom. Peirce writes: "I may mention, for the benefit of those who are curious in studying mental biographies, that I was born and reared in the neighborhood of Concord – I mean in Cambridge – at the time when Emerson, Hedge, and their friends were disseminating the ideas that they had caught from Schelling, and Schelling from Plotinus, from Boehm, or from God knows what minds stricken with the monstrous mysticism of the East. […] and I am not conscious of having contracted any of that virus. Nevertheless, it is probable that some cultured bacilli, some benignant form of the disease was implanted in my soul, unawares, and that now, after long incubation, it comes to the surface, modified by mathematical conceptions and by training in physical investigations." Peirce 1931–35: VI. 103.

4 Pavlov developed a more sophisticated model in his further work, and the neo-behaviorists like Osgood postulated an inner space of semantic dispositions which direct the creation of a semantic network (cf. Osgood, 1976).

if Lorenz quickly uttered sounds and the freshly hatched out goose registered his presence, he filled the slot of a "mother-goose". As soon as this concepts had been fixed, he remained the prototype of a "mother-goose" for them. Lorenz generalized this observation and postulated so-called "super releasers", i.e., very primitive schemata which only ask for minimal perceptual and neural control. They may even fit better to formal molds than to biologically real entities. In the process of "Prägung", a rich semantic system is developed which could not have been coded genetically. The super-releaser has an optimal effect but minimal conditions for its application.

c) A similar concept of "Bedeutungswelt" (meaningful universe) was proposed by another biologist, Jakob von Uexküll. His theoretical biology influenced Cassirer's philosophy of symbolic forms, while they were both teaching in Hamburg. For Uexküll, every animal creates its own "Bedeutungswelt", which depends first on its windows of perception and then on its vital needs. Thus a tick reacts perceptually to the concentration of butyric acid, typical for warm-blooded animals. If this stimulus is received, the tick drops and eventually gets into the fur of the animal. As it feeds on the blood of the animal, this minimal semiotic system guarantees its survival.

d) In the realm of psychology, Gibson elaborated the concepts of "prägnante Gestalt" and "Valenz" of his teacher Koffka and coined the term of *affordance*. Any object or process in our environment may have affordances; thus a chair allows for sitting, a bed for sleeping, etc. Therefore, the whole ecology is meaningful in terms of possible actions or events it allows (it signifies the potential events and actions). As some actions are more frequent or have more value for survival than others, the concept of affordance may be graded. Maximal affordance would then correspond to "Prägnanz".

e) Cassirer generalized the perceptual/motoric "Prägnanz" to "symbolische Prägnanz" as perception is the first level of semiosis (~ symbol creation) in his system. If for Gestalt-psychology meaningfulness was one of the criteria of a good gestalt, for Cassirer meaning, i.e., symbolic content, is the precondition of true perception, insofar as elements of perceptual input have to be integrated into a system of perceptually based concepts, which is created by the mind. The signs associated with these concepts allow the halting of the steady stream of consciousness, in which something stable is fixed and brought to attention.[5]

f) The American psychologist David Hebb, whose work has become basic for modern neuropsychology (cf. Hebb-assemblies), discusses different types of instincts and two models. The first model, which may be called monogenetic, assumes one basic force field of biological motivation which is then separated into different subfields like: hunger, thirst, sexual appetence, aggression, fear, intellectual curiosity, etc. The second assumes the preexistence of a topology of motivations with different centers (attractors); cf. Hebb (1959: 246 f.).

g) The application of the concept "Prägnanz" (imprinting) to language acquisition was proposed by William Stern, who recognized a type of goal-oriented, internally controlled process in language acquisition. Thomae linked this idea with the more general discussion in ethnology and psychoanalysis (cf. Thomae 1972: 244)[6] Freud's theory of psycho-genital development assumes far-reaching consequences of basic processes in early childhood, which fix specific personality traits for the whole life of a person. Thomae included also the concept of "canalization" which assumes a primary morphogenesis of the epigenetic field, in which all further developments are inscribed. This idea links the concept of "Prägnanz" with the discussion between Waddington and Thom on epigenetic fields in the 60s and thus prefigures Thom's semiophysics of "prégnance".

Summarizing this discussion which has almost occupied biologists and psychologists for more than one century, one can assume that on one side all animals have specific windows to their ecology and on the other side that a small set of motivations or vital force fields exist. The list of relevant basic types remains an open question. René Thom just presupposes the existence of such a basic and very simple set, which shares the feature of structural stability and is independent from specific material contexts. This basic set is subdivided into "salience" and "pregnancy".

---

5 Cf. Cassirer 1923/2001: 20 and van Heusden 2003: 126 f.

6 Bischof (1989) replaced the concept of a sensitive period of imprinting by the concept of a "sensitive situation". Such situations may even occur later in life, e.g., after sexual maturity in humans (and mammals). He proposed to separate "releasers" based on perception and "detectors" linked to behavioral systems ("Funktionskreis"). The instinctive behavior described by Lorenz belongs to a biologically necessary (releaser) imprinting, whereas sexual imprintings belong to behavioral systems using detectors (cf. Bischof, 1994).

To these rather general ideas already found in the discourse of the contemporary scientific community, Thom adds two new ideas:

- The salience effect in perception may be linked via psychophysical laws to the dynamics of objective fields in physics and chemistry; i.e., perception is related in its basic principles to laws in natural science. In Thom's terms, there is an explanatory continuum between objective salience in physics and chemistry and subjective salience in cognition and semiosis.
- The topology of salient objects and events governs the flow of pregnancies via a process called "diffusion de prégnance" (channeling of pregnancy). As in Lorenz's case of "Prägung" (imprinting), the attractor landscape of pregnancy is filled with actual or memorized salience effects. In this channeling, the multiple forms of (perceptual) salience are molded into the instinctive and inborn forms and elaborate them to rich and context-dependent fields of categorical perception and behavior[7]

Thom's hypotheses apply mathematical techniques and theorems to solve the most basic problems. But the psychophysical transition calls for the application of the laws of physical dynamics, not only the mathematics applied in this field. This line of thought has been elaborated by the group of Turvey, Kelso and others (cf. Kelso, 1997, and for linguistic applications Wildgen, 1994: chapter 3). The diffusion of pregnancy calls for something like fluid dynamics and models of growth in space and time. The mathematical and physical aspects of Thom's proposals have been discussed and elaborated in Wildgen (1982, 1985, 1994 and 2005), and we shall concentrate in the following on the philosophical and semiotic aspects of Thom's proposals. The work of Cassirer (mainly after 1935) will guide our philosophical inquiry.

7 In Petitot-Cocorda (1992) this idea is elaborated using the term "physics of meaning" instead of "semiophysics". In chapter V, 3 he introduces more advanced mathematical tools and explains fractal structures and chaos. The topics "prégnance" and "saillance" are discussed in chapter VI, 5 (p. 315–324).

4. Mathematics applied to psychology and semiotics: The objections by Cassirer

Cassirer discussed in several papers the application of Klein's theory of invariants (the Erlangen Program) to the psychology of perception.[8] The basic question, which has been asked again in cognitive semantics (cf. Lakoff, 1987: Chapter 14) and in Leyton's critique of the Erlangen program (cf. Leyton, 2001), is: How is the mathematical study of the physical world related to the study of the mental and the cultural word? Is there some kind of continuity such that the concepts of mathematics that are successful in the scientific capture of features in the objective world may be applied (with the same success) to the mental and the cultural world? In mathematical Platonism (favored by many mathematicians, e.g., René Thom in his early publications), the mathematical model has an existence in its own right, and its application to physics is just one possible field among many. A more synthetic view of mathematics assumes a specific adaptation to major problems of physics and technology in the course of the millennia. Kepler, Galileo, later Newton, Leibniz, and Einstein took profit of the conceptual potential of mathematics (geometry, trigonometry, analysis and non-Euclidean geometry) to put forward advanced theories of the universe and of basic physical processes. Leyton (2001) argues that the invariant structures proposed by Felix Klein as universal explanatory devices eliminate the information contained in visual and linguistic (symbolic) forms and are therefore insufficient.[9] Cassirer argued in several articles written towards the end of his life that mathematics (mainly Klein's geometry and topology) describe only a realm of possibilities (e.g., possible Euclidean and non-Euclidean geometries) and that an empirical study of visual perception can at its best select, specify, even combine these geometries or their features. The basic selection must be made based on empirical arguments. The fact that certain mathematical

8 Felix Klein had developed the main ideas together with the Norwegian mathematician Sophus Lie, who specialized in continuous groups. When Klein became Professor in Erlangen in 1872, he formulated the program of a generalized geometry in his opening lecture (cf. Klein, 1872/1974).

9 The more relevant process is that of deformation and the transformations which are responsible. This view is compatible with Chomsky's transformational analysis, which was criticised as "objectivistic" by Lakoff. Lakoff and Núnez (2000) argue that mathematics are a specialization and a deformation of natural (language) reasoning and are therefore biased in view of an analysis of language and culture.

structures have been invented historically and successfully applied in physics and chemistry cannot predict the success of their application to perception and language. This skeptical judgment by a philosopher who since his time as a student in Marburg had followed the development of natural sciences and mathematics should direct our evaluation of Thom's project of "semiophysics". Although his approach is anti-reductionist and strictly opposed to physicalism, it is very optimistic as to the relevance of topological dynamics and dynamic system theories for the sciences of man and society.

Leyton's criticism of the Erlangen program affects Thom's models insofar as elementary catastrophes correspond to the geometrical groups of regular polygons, Platonic solids, etc. (cf. Slodowy 1995).[10] Therefore, one may assume that the elementary catastrophes are structures stripped off their (semantic) information and that via deformation, breaking of symmetry, and combination of the catastrophe schemata the information lost may be reconstructed. In this perspective, the major concern is not how to find (or construct) the archetypes, but to describe the modes of accumulating meaning, starting from such archetypal but meaningless forms. The process of "diffusion of pregnancy" becomes a process of meaning *creation* rather than of meaning *distribution* from an original source full of (pregnant with) meaning.

5. The propagation of "pregnancy" (basic meaning) and the origin of language

Thom's theory of propagation of basic meaning is essentially a causal theory of meaning. All basic meanings are "theoretical entities", i.e., they are difficult or impossible to observe, but must be assumed in order to understand the observable effects. We may observe basic behavioral categories in newborn children (first weeks and months), register the semantic structures in one-word or two-word utterances and finally describe the lexicon and syntax of the adult language. These stages form a causal chain with four major strata:

10 The cuspoïds correspond to the series of regular polygons, the umbilics to the set of dihedra (having two differently directed faces), and the series E6, E7, E8 corresponds to three classes of Platonic solids (the five regular polyhedra minus their duals).

S_0	$> S_1$	$> S_2$	$>S_3$
before birth	newborn child, prelinguistic	linguistic start	adult competence

Figure 1. Four stages in language development.

The theoretical starting level S_0 must evolve regularly given a human genome. At birth only some auditory capacities already developed in the womb exist; the first period of development concerns other sensory organs and basic motor programs (cf. for Piaget's sensory-motor level of cognitive development Piaget 1973: chapter 3). The genome cannot code for specific sensory forms or motor-patterns, i.e., it can only fix some gradients that allow the detection of relevant input in order to trigger the process of elaboration (intrinsic) and learning (extrinsic). Parallel to these basic gradient-fields, the sensory capacities and corresponding categorical and memory capacities evolve. For simplicity's sake, I assume only *one* basic (inborn) gradient (a correlate should be found in the genetic code) called P (pregnancy). As the sensory capacities S_i (saliences) evolve, P is interpreted by the content of S_i, via a regular fitting between P and S_i. As a precondition of such a fitting, our sensory organs must be in a rough harmony with P. A first division of P may be due to the multiplicity of S_i. If the different sense organs evolve in a consecutive order, e.g., $S_{au(ditive)} \rightarrow S_{vi(sual)}$, the types of P invested by S_{au} may condition the structure of P invested by S_{vi}. This leads to a cascade of P_k invested by S_i:

P invested by S_{au}	$\rightarrow$	P_k (S_{au})
P invested by S_{vi}	$\rightarrow$	P_k (S_{vi})

Figure 2. Cascade of investments.

The same thing may occur with all sensory fields (whose number is open, traditionally five, maximally 30). If the first investments prefigure the later, we obtain a more complicated net:

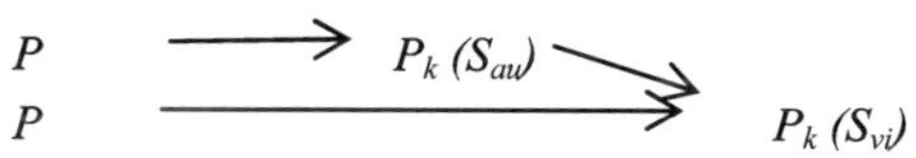

Figure 3. Network of investments.

So far the propagation described a general rule in the formation of sensory schemata in animals. I have neglected the motor aspect for which a parallel process of propagation may be assumed. Both are further coordinated as in the case of the perception of a predator, followed either by flight or a mobbing action.

A new level is reached in *alarm calls* or other socially relevant and culturally transmitted signals. In these cases, perception is coupled to a behavior which is able to redistribute the perception and to trigger collective action in a community (independent of consciousness, free will or intentionality).[11] In this case, the category of Interpretant (a lawful and stable link between sign and meaning) begins to surface in behavior (not necessarily in consciousness). If the individual formation of a pregnant sign P, which links perception and action, is the base line, then the alarm is efficient by virtue of the stability of the link between both relata "Perception" and "Object" (in Figure 6) across the community of adults. This enables true thirdness (involving the Interpretant) in terms of Peirce's sign classification.[12]

The alarm-call is perceived and causes action (in adult animals who have learned the rule).[13] The cascade of semiosis can be further completed if a semiotic culture, i.e. a stable and shared system of signs, which governs social life, evolves. This seems not to be the case for most primates, although it is difficult to detect animal sign systems for humans if they do not have clear cut ecological or behavioral consequences (if they don't fit a stimulus-response schema).

---

11 Cf. Wildgen (2004) for the evolution of language and the role of dynamic patterns in it. In Bax, van Heusden and Wildgen (2004), some aspects of biological and cultural evolution are discussed. In Wildgen (forthcoming 2010: Chapter 13) the field of neuro- and biolinguistics is evaluated.

12 Beyond a stable correlation, a complexity reduction must occur, i.e. the interpretant must be less complex than the totality of dyadic relations encountered or present in the environment. This topic is discussed in Luhmann (2004) in relation to social systems and measures of complexity proposed in cybernetics.

13 Basically our simplified models only focus on the relations to the object (R–O and I–O). The iteration opens an infinite regress, which may be halted by levels of attention or consciousness. The infinite series may be compressed by integration because the process of thought is basically a continuous one, which is only segmented artificially for the sake of reconstruction. The underlying paradox is similar to the classical paradox of Zenon (the race between Achilles and the turtle).

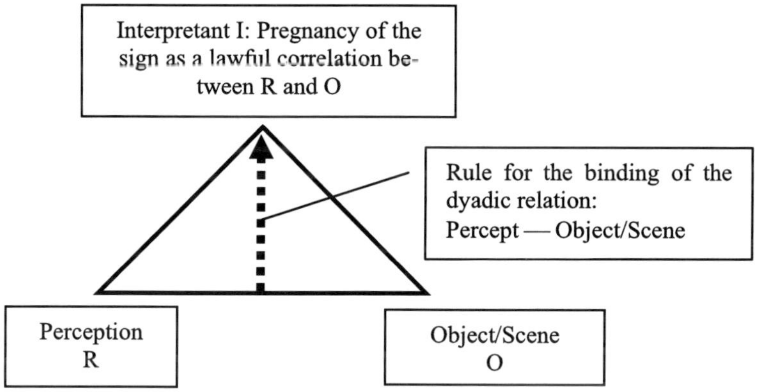

Figure 4. The „alarm-call“ as an „Interpretant“.

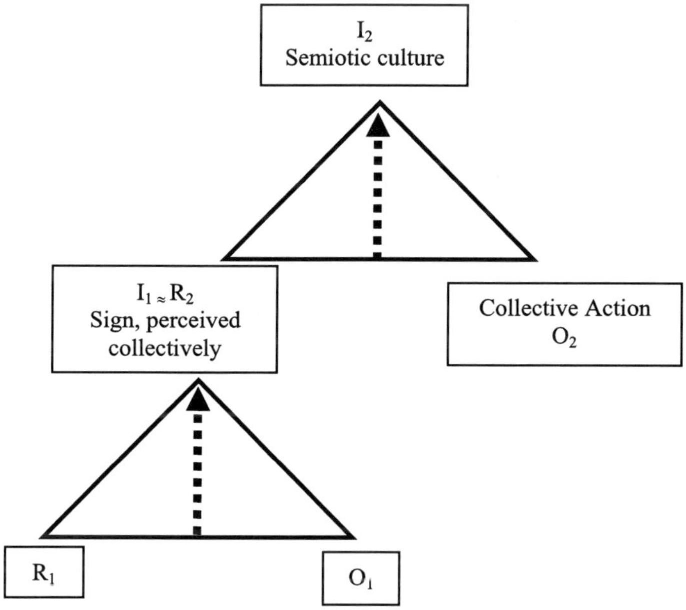

Figure 5. The second level and the start of an iterative process of semiosis.

The semiotic cascade can be continued if social intelligence and a high level of consciousness (theory of mind, mind-reading) enables the formation of a system of collective knowledge, which is learned and stabilized (adapted) via a system of signs. The new object of the sign (its referent) is now the collective knowledge (or sections of it).

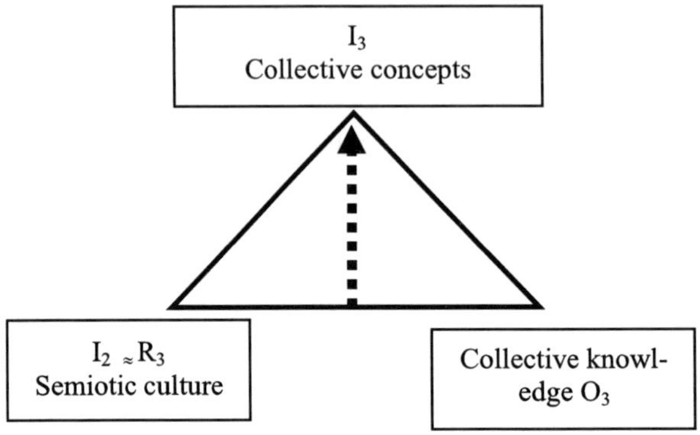

Figure 6. The third level of semiosis (the first is not represented).

Under these conditions the origin of language and linguistically based cognition may be modeled in analogy with the origin of life, i.e., the transition from simple chemical structures to living beings. The key to the transition is a type of self-organization based on hypercycles (cf. Eigen and Schuster, 1979 and applied to semiotics Wildgen, 2007). I shall try to comment this idea, because bifurcation alone seems to be a mechanism too poor and unstable to explain the emergence of such a powerful faculty as the faculty of symbolic behavior and language are.

In a first period the chain $I_1 \rightarrow I_2$ just allows for a powerful distribution, i.e., the stable perceptual reaction of one or a few animals reaches the whole population via signs. In a second period this chain contributes to socialize the individual perception-action cycles. In such a cycle, "interaction, exploration, and information are co-specific (or co-implicative)" as Turvey, Carello, and Kim (1995: 282) tell us. As a consequence, a network in which information circulates is created; individual problem solving is diffused through social learning and thus a new level of competence is enhanced in the community. The coordination of individual perception/action cycles allows for a (social) control on the performance of others, which again stabilizes the patterns of collective action and creates a stable distribution of roles (in chasing, production, child-rearing etc.). In the sequel, social intelligence and social competence are further developed.[14]

The binding of signs and action patterns at level I_2 (semiotic culture) involves mechanisms of social control, by which the result of an action is anticipated in view of a collective goal (e.g. the specific shape of a hand axe) and the result evaluated relative to the goal (the ideal). In the process of instruction (and education) chains of such control processes on action/perception cycles occur, the ideal (good) forms of artifacts, rituals, and behavior in general establish a higher control to which also teachers must submit. These considerations show that cyclic and self-referential structures exist already at the single levels; therefore the cycle which links the levels I_1, I_2 and I_3 in a self-referential mode can be called a *hypercycle* in the sense used by Eigen and Schuster (1979).[15] Figure 9 at left describes the simple process of pregnancy diffusion described by Thom (the length of the chain and possible interferences by imitation and learning are neglected). The right part of the figure gives a schematic picture of the hypercycle which links the three levels of Interpretants in a self-referential fashion.

14 In Herrmann et al. (2007) the evolution of social cognition is described as a major motor of human evolution.

15 Eigen and Schuster (1979: 3f) mention as examples of cycles: the enzyme which turns a substrate into a product, the carbon-cycle proposed by Bethe-Weizäcker, "which keeps the sun shining" (ibidem), the Krebs-cycle, which regulates the metabolism of the living cell. Their main examples of hyper-eycles are the precursors of the reproduction and translation apparatus found in present prokaryotic cells, and thus the basic mechanism underlying the origin of life (ibidem: 62). The Darwinian nature of hyper-cycles is described in the following passage: "Non-coupled self-replicative units guarantee the conservation of a limited amount of information which can be passed on from generation to generation. This proves to be one of the necessary prerequisites of Darwinian behaviour, i.e. of selection and evolution [...]. In a similar way, catalytic hyper-cycles are also selective, but in addition, they have integrating properties, which allow for cooperation among otherwise competitive units. Yet they compete even more violently than Darwinian species with any replicative entity not being part of their own. Furthermore, they have the ability of establishing global forms of organization as a consequence of their once-forever-selection behaviour, which does not permit a coexistence with other hypercyclic systems, unless these are stabilized by higher order linkage." (ibidem: 6).

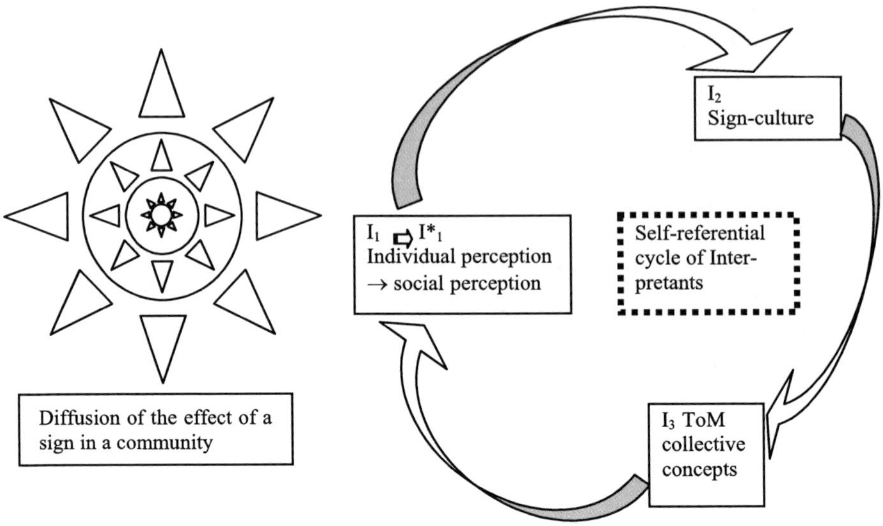

Figure 7. Diffusion of a sign (left) and elaboration of the chain: I1→ I2 → I3 with closure in I*1.

The closure means that the categories transported and possibly modified in the chain $I_1 \rightarrow I_2 \rightarrow I_3$ have again a determining effect on the primary (perceptually based) semiosis I_1. As a consequence the individual mind begins to react to socially stabilized and selected categories ($I_3 \rightarrow I_1$). In terms of the Sapir-Whorf hypothesis, the cultural word-view (in this case produced by selective transformations in the chain $I_2 \rightarrow I_3$) modifies I_1 and produces I_1* which again shapes the signs at the next level I_2* and so forth. The cycle tends to be cooperative insofar as all the elements in the cycle adapt to form a collective mode, and it is violently competitive against any replicative structure not being part of the cycle (cf. Eigen and Schuster, 1979: 6). I shall not elaborate this topic, but it is clear that the diffusion of pregnancy on a single level is not sufficient to explain the dramatic shift between animal communication and human language. The question is: Can such a three level semiosis and a self-referential hypercycle explain the emergence of grammar and of syntactic complexity?

6. The emergence of syntactic complexity

In his first period of biological and semiotic theorizing, Thom (1972) had established a list of archetypical schemata and made the conjecture that they are underlying the linguistic valency patterns described by Tesnière (1959).[16] The question is: How are the archetypes proposed by Thom in 1972 linked to the phenomena of pregnancy and pregnancy diffusion, which were put forward in Thom 1988?

The simple propagation of "pregnancy" produces maximally a series of attractors with diminishing energy. Each bifurcation diminishes the energy of the single attractors. The dynamic systems with 2, 3, 4 … attractors correspond to the cuspoïds A_3, A_5, A_7,… in Thom's classification theorem. Thom describes the valency patterns as conflict-scenarios of pregnancy and bifurcations in the diffusion of pregnancy. In order to specify the set of archetypes, one needs however a richer basis:

a) The geometry of the umbilics D_4, D_5 (and of the compactified E-series; cf. Wildgen, 1982: 85–92).

b) A model of dominance and dominance shift. Thom mentions two types of dominance dynamics: perfect delay and thermodynamic shift (Maxwell convention); an explicit model would ask for stochastic dynamics and thus for the enlargement of the mathematical basis (synergetics, chaos theory, cf. Wildgen and Mottron, 1987, Petitot-Cocorda, 1992, Wildgen, 1994 and 1998).

16 The notion of valency can be traced back to similar notions in the classical Greek (Thrax) and Indian grammar (Panini). Tesnière has put it programmatically into the centre of his "structural syntax". Several European schools have taken it up in the seventies and eighties (Helbig, Greimas, Mel'čuk). Formalization and computer models based on valency have been designed in the United States in the late sixties and seventies (Gaifman) and most modern grammars like LFG (lexical functional grammar), HPSG (head driven phrase structure grammar), GB (government and binding) make use of the concept of valency. Some authors, e.g. Mel'čuk, distinguish between morphological, syntactic and semantic valency (cf. Schneider 1996). The notion Thom elaborates (he just refers to Tesnière and not to the sequel) is nearer to *semantic* valency than to *syntactic* valency. As Lamb (2002) shows, the semantic (sememic) stratum is connected to the perceptual centres in the brain, and therefore semantic valency has to do with the way we perform and perceive basic actions. They are probably formed before the linguistic means to express them are acquired. Therefore semantic valency is located in a domain intermediate between cognitive (perceptually and behaviourally based) schemata and structures of meaning.

c) A classification of all relevant paths in the bifurcation plan (i.e., not just the analysis in the neighborhood of the singularity in some simple cases).

The second need concerns the technique of enrichment (attributional elaboration) of the formal archetypes. I shall develop some ideas in order to show how to fill this lacuna (based on my proposals in the last chapter). It seems plausible to see the valency patterns emerge in social interaction based on a theory of mind (a representation of the other and his typical action in social encounters). In the model sketched above, the self-referential loop $I_1 \rightarrow I_2 \rightarrow I_3 \approx I_1^*$ triggers the categorization of basic types of action and social interaction, which remain implicit at the levels I_1 and I_2, but are systematically encoded at level I_3. The socially distributed knowledge encompasses: simple patterns of locomotion (with a source and a goal), actions on objects, on agents (other minds), exchange and distribution of objects in a group, interpersonal conflicts and alliances, etc. These types of social contents are at the same time perceived (I_1^*) and then again managed with the use of signs (I_2^*). The hypercycle selects the stable and recurring features of social action and communication that control the communicative (emotional, political, economic) relations in the group. This selective abstraction establishes a new level I_3 which may be called theory of *social* mind. In the transmission of structure from I_3 to I_{1*} (perception) and I_{2*} (signs), the third level categories finally become cognitive schemata, or to use a term coined by Fillmore: semantic roles, frames, scenarios (cf. Wildgen, 1985: chapter 1 for a review of this tradition). The hypercycle thus incorporates selected scenarios of action/interaction into semantic-syntactic schemata which organize sentences.

The pregnancy is different at the levels I_1 (perception), I_2 (signs), and I_3 (incorporated scenarios) but the topologico-dynamic invariants put forward by Thom guarantee the coherence of the three levels and imprint a topologico-dynamic structure on linguistic frames.[17] Language becomes in the consequence of this evolution independent from hic-et-nunc situations, it allows for the decontextualization of human communication and thus distin-

17 In Wildgen (1994: chapter 5) four major domains of interpretation were distinguished: locomotion in space, change in a quality space, external action/interaction and internal action. The dynamic schemata are common to all domains. It is clear that evolution proceeds with minimal steps and that the three levels distinguished in Figure 10 (right side) are plateaus of rather small changes after a period of rapid changes. The number of these plateaus may be multiplied or diminished.

guishes humans from (most) animals.[18] The exact format of such a model with self-referential organization must be further developed and cannot be fully specified in this contribution.

7. Conclusion

The great merit of René Thom's thinking (since the mid-sixties) lies in his focus on very basic questions of biology, semiotics, and linguistics. But it was not sufficient, as Tesnière (Thom's major linguistic inspiration) did, to point to analogies between structural syntax on one side and chemistry (valency) and drama constellations (actants) on the other. These hints (already Peirce had conjectured such a link) had to be specified. In a first move, Thom proposed a mathematically based conjecture on the nature of valency patterns based on the classification of very general process-types (derived from catastrophe theory). The semantic archetypes (cf. Wildgen, 1982, 1985) are pure theoretical entities, they formulate a program of research that gives rise to empirical analyses, such that some of them may be elaborated or falsified in later work. In the late seventies, Thom took up the notion of "prégnance" with its multiple readings in different disciplines (biology, etiology, psychology, linguistics). His proposals for an evolutionary process remained in the intellectual horizon of epigenetic models and should be brought nearer to current thinking in evolutionary theory (which is basically Darwinian and not morphological). I think that both lines of thought can be unified, insofar as Darwinian evolution does not preclude the effect of internal factors of coherence and stable epigenesis. The concept of hypercycle introduced by Eigen and Schuster (1979) is a possible bridge because it is firmly rooted in Darwinian premises. They state (ibidem: VII):

> If we analyze the conditions of hypercyclic organization we immediately see their equivalence to the prerequisites of Darwinian selection. The latter is based on self-reproduction which is a kind of linear autocatalysis. The hypercycle is the next higher level

18 Donald (2001: 136f) argues that symbolic communication is in spite of the abstractness of self-constructions and symbolic systems always rooted in physical space and physical fields (forces). At this point he adopts the main arguments on embodiment, which Lakoff and Johnson (1980) had reanimated based on Husserl's and Merleau-Ponty's phenomenology. In Petitot-Cocorda (1992) a more systematic reading of Husserl's phenomenology, without the scepticism of Lakoff, is put forward.

> in a hierarchy of autocatalytic systems [...]. It is made up of autocatalysts or reproduction cycles which are linked by cyclic catalysis, i.e. by another superimposed autocatalysis. Hence a hypercycle is based on non-linear (e.g. second or higher order) autocatalysis.

The further elaboration of the model would have to consider the catalytic properties already at the levels I_1 and I_2, i.e. already perception is made up from cyclic processes in which memory and actual inputs interact, and the same is probably true for signal transmission and reception (and the necessary ongoing learning processes). Compared to the explicit treatment of chemical autocatalysis in the formation of RNA by Eigen and Schuster (1979), there is however still a long way to make for semiotic theorizing.

References

Bax, Marcel, Barend van Heusden and Wolfgang Wildgen (eds.) (2004). *Semiotic Evolution and the Dynamics of Culture*, Peter Lang, Bern.

Bischof, Norbert (1994). *Das Rätsel Ödipus: die biologischen Wurzeln des Urkonfliktes von Intimität*, 4. Aufl., Piper, München.

Brandt, Per Aage (1992). *La charpente modale du sens. Pour une sémio-linguistique morphogénétique et dynamique*, Aarhus U.P., Aarhus.

Cassirer, Ernst (1923, 1925; 1929). *The Philosophy of Symbolic Forms*. Vol. 1: Language; vol. 2: Mythical Thought; vol. 3: The Phenomenology of Knowledge, Yale UP, New Haven & London: 1955–1957.

Cassirer, Ernst (1950). *The Problem of Knowledge. Philosophy, Science, and History since Hegel.* Translated by W.H. Woglom and C.W. Hendel. With a Preface by Charles W. Hendel, New Haven and London.

Donald, Merlin (2001). *A Mind so Rare: the Evolution of Human Consciousness*, Norton, New York.

Eigen, Manfred and Peter Schuster (1979). *Hypercycle. A Principle of Self-Organization*, Springer, Berlin.

Fillmore, Charles J. (1977). "Scenes-and-Frames Semantics", in: A. Zampolli (ed.), *Linguistic Structures Processing*, North Holland, Amsterdam: 55–81.

Gaifman, H. (1965). "Dependency Systems and Phrase Structure Systems", in: *Information and Control* 8: 304–337.

Greimas, Algirdas Julien (1966). *Sémantique structurale. Recherche de méthode*, Larousse, Paris.

Guckenheimer, H. et Ph. Holmes (1983). *Nonlinear Oscillations, Dynamical Systems, and Bifurcations of Vector Fields*, Springer, Berlin.

Haken, Herman and Michael Stadler (eds.) (1990). *Synergetics of Cognition. Proceedings of the International Symposium at Schloß Elmau,* 1989, Springer, Berlin.

Haken, Hermann (1996). *Principles of Brain Functioning. A Synergetic Approach to Brain Activity, Behavior and Cognition*, Springer, Berlin.

Hebb, Donald Olding (1959/1966). *A Textbook of Psychology*, (new edition 1966), Saunders, Philadelphia.

Hermann, E. et al. (2007). Humans Have Evolved Specialized Skills of Social Cognition: The Cultural Intelligence, in: *Science* 317: 1360–1366.

Heusden van, Barend (2003). "Cassirers Ariadnefaden – Anthropologie und Semiotik", in: Hans-Jörg Sandkühler and Detlev Pätzold (eds.) (2003). *Kultur und Symbol. Ein Handbuch zur Philosphie Ernst Cassirers*, Chapter 5: 111–147.

Kelso, J. A. Scott (1997). Dynamic Patterns. The Self-Organization of Brain and Behavior, Bradford, Cambridge (Mass.).

Klein, Felix (1872/1974). *Das Erlanger Programm. Vergleichende Betrachtungen über neuere geometrische Forschungen*, Geest & Porzig, Leipzig.

Lakoff, George and Mark Johnson (1980). *Metaphors We Live By*, Chicago University of Chicago Press Chicago.

Lakoff, George and Rafael E. Núnez (2000). *Where Mathematics Comes From – How the Embodied Mind brings Mathematics into Being*, Basic Books, New York.

Lamb, Sidney (2002). "Learning Syntax – A Neurolinguistic System", reprinted in: Lamb, Sidney (2004). *Language and Reality*, Continuum, London: chapter 15.

Leyton, Michael (2001). *A Generative Theory of Shape*, Springer, Heidelberg.

Luhmann, Niklas (2004). *Einführung in die Systemtheorie* (ed. by Dirk Baecker), 2nd edition, Heidelberg, Carl-Auer-Systeme.

Mel'čuk, Igor A. (1988). *Dependency Syntax. Theory and Practice*, State University of New York Press, New York.

Mithen, Steven (1998). "A Creative Explosion. Theory of Mind, language and the Disembodied Mind in the Upper Palaeolithic", in Idem (ed.) (1998). *Creativity in Human Evolution and Prehistory*, Routledge, London: 165–191).

Nicolis, Grégoire and Ilya Prigogine (1989). *Exploring Complexity. An Introduction*, Freeman, New York.

Osgood, Charles (1976). *Focus on Meaning*, Mouton, The Hague.

Peirce, Charles Sanders (1931–35). *Collected Papers of Charles Sanders Peirce* (ed. by Charles Hartshorne and Paul Weiss), Harvard U.P., Cambridge (Mass.), vol. I–VI (cited with vol. and §).

Petitot-Cocorda, Jean (1985). *Morphogenèse du sens*, Presses Universitaires de France, Paris (English translation: (2003). Morphogenesis of Meaning, Lang, Bern).

Petitot-Cocorda, Jean (ed.) (1988). Logos et Théorie des Catastrophes, Patiño, Genf.

Petitot-Cocorda, Jean (1992). *Physique du sens. De la théorie des singularités aux structures sémio-narratives*, Éditions du CNRS, Paris.

Piaget, Jean (1973). *The child and reality. Problems of genetic psychology (Problèmes de psychologie genétique)* translated. by Arnold Rosin), Grossman, New York.

Schneider, Gerold (1996). *A Linguistic Comparison of Constituency, Dependency and Link-Grammar*, Lizenziatsarbeit, Zürich. Internet site visited 26/6/2009: http://www.ifi.uzh.ch/cl/study/lizarbeiten/lizgerold.pdf.

Slodowy, Peter (1988). "Platonic Solids, Kleinian Singularities, Elementary Catastrophes, and Lie Groups", in: Jean Petitot-Cocorda (ed.) *Logos et théorie des catastrophes. A partir de l'œuvre de René Thom*, Patiño, Geneva: 73–98.

Sussmann, Hector J. and Raphael S. Zahler (1978). "Catastrophe Theory as Applied to the Social and Biological Sciences: A Critique", in: *Synthèse* 37, 117–216.

Tesnière, Lucien (1959). Élements de syntaxe structurale, Klinksieck, Paris.

Thom, René (1972/1977). *Stabilité structurelle et morphogenèse. Essai d'une théorie générale des modèles* (2nd edition, revised and augmented 1977), InterEditions, Paris, (citations are based on the second edition).

Thom, René (1975). *Structural Stability and Morphogenesis*, Benjamin, Reading: (translation with some omissions from the French original: Stabilité structurelle et morphogenèse 1972).

Thom, René (1988). *Esquisse d'une sémiophysique. Physique aristotélicienne et théorie des catastrophes*, InterEditions, Paris (translation into English: 1990. Semiophysics: A Sketch, Addison-Wesley, Redwood City, Calif.).

Thom, René (1990). *Apologie du Logos*, Hachette, Paris.

Thom, René (1991/2003). "Saillance et Prégnance", first published in: *L'inconscient et la Science*: 64–82, (ed. by R. Dorey), Dunod, Paris; citations are based on Thom, 2003.

Thom, René (2003). *Oeuvres complètes*, CD-Rom, Institut des Hautes Études Scientifiques, Bures-sur-Yvette (Paris).

Thomae, H. (1972) "Entwicklung und Prägung", in: Idem. (ed.) *Entwicklungspsychologie.* Handbuch der Psychologie in 12 Bd., Hogrefe, Göttingen: 240–311.

Turvey, M.T., C. Carello, and Nam-Gyoon Kim (1990), "Links between Active Perception and the Control of Action", in: Haken and Stadler (1990: 269–295).

Uexküll, Th. v. (1984). "Zeichen und Realität als anthroposemiotisches Problem", in: Oehler, K. (Hg.), *Zeichen und Realität*, Narr, Tübingen.

Wildgen, Wolfgang (1979). *Verständigungsdynamik. Bausteine für ein dynamisches Sprachmodell*, thesis of Habilitation, University of Regensburg; electronic copy: <http://elib.suub.uni-bremen.de/ip/docs/00010402.pdf>.

Wildgen, Wolfgang (1981). „Archetypal Dynamics in Word Semantics. An Application of Catastrophe Theory", in: Hans-Jürgen Eikmeyer and Hannes Rieser (eds.). *Words, Worlds and Contexts. New Approaches to Word Semantics*, de Gruyter, Berlin: 234–296.

Wildgen, Wolfgang (1982). *Catastrophe Theoretic Semantics. An Elaboration and Application of René Thom's Theory*, Benjamins, Amsterdam.

Wildgen, Wolfgang (1985). *Archetypensemantik. Grundlagen für eine dynamische Semantik auf der Basis der Katastrophentheorie*, Narr, Tübingen.

Wildgen, Wolfgang (1994). *Process, Image, and Meaning. A Realistic Model of the Meanings of Sentences and Narrative Texts*, Benjamins, Amsterdam.

Wildgen, Wolfgang (1998). "Chaos, Fractals and Dissipative Structures in Language. Or the End of Linguistic Structuralism", in: Gabriel Altmann and Walter A. Koch (eds.). *Systems. New Paradigms for the Human Sciences*, de Gruyter, Berlin: 596–620.

Wildgen, Wolfgang (1999). *De la grammaire au discours. Une approche morphodynamique*, Lang, Bern.

Wildgen, Wolfgang (2004). *The Evolution of Human Languages. Scenarios, Principles and Cultural Evolution*, Benjamins, Amsterdam.

Wildgen, Wolfgang (forthcoming 2010). Die Sprachwissenschaft des 20. Jh.s: Versuch einer Bilanz, Berlin, De Gruyter.

Wolfgang Wildgen und Laurent Mottron (1987). *Dynamische Sprachtheorie. Sprachbeschreibung und Spracherklärung nach den Prinzipien der Selbstorganisation und der Morphogenese*, Brockmeyer, Bochum.

Zeeman, E.C. (1977). *Catastrophe Theory: Selected Papers 1972–1977*, Addison-Wesley, Cambridge (Mass.).

Urban Universals[♣]

ISABEL MARCOS

1. Introduction

The subject of my article will be to enumerate the three consequences that we can draw from the following phrase:

I believe that the city is not, and should not be, an amorphous space where the diverse architectural interventions float like a whole lot of islands. The city is, as I have demonstrated by my doctoral research (1996), a dynamic topology which is deployed in history.

Following that, within the body of the article, I shall explicate at length the latest inferences drawn from the above phrase. On this subject, I shall analyse an "iluminura (illuminated picture)" dating from the times of King *Dom Manuel* (1495–1521). Indeed, the reign of this king marked a key moment, that of the structuring of a new style of society, the Great Discoveries being without doubt the starting event. The *iluminura* chosen embodies this epoch and through its pictorial devices represents a dynamic topology which deploys diverse temporalities.

1st Inference

From a theoretical point of view, the perspective offered enables one to think out the city form through an evolving cartographic representation in three strata (*cf.* Figure 1).

♣ This article has been made possible thanks to the support of the Foundation of Science and Technology of Portugal and the e-GEO – the Centre of Geographical and Regional Development Studies – Universidade Nova de Lisboa. The diagrams were conceived and realised by Isabel Marcos and Catarina Bello. This article was translated from the original text in French by Ian MacCabe.

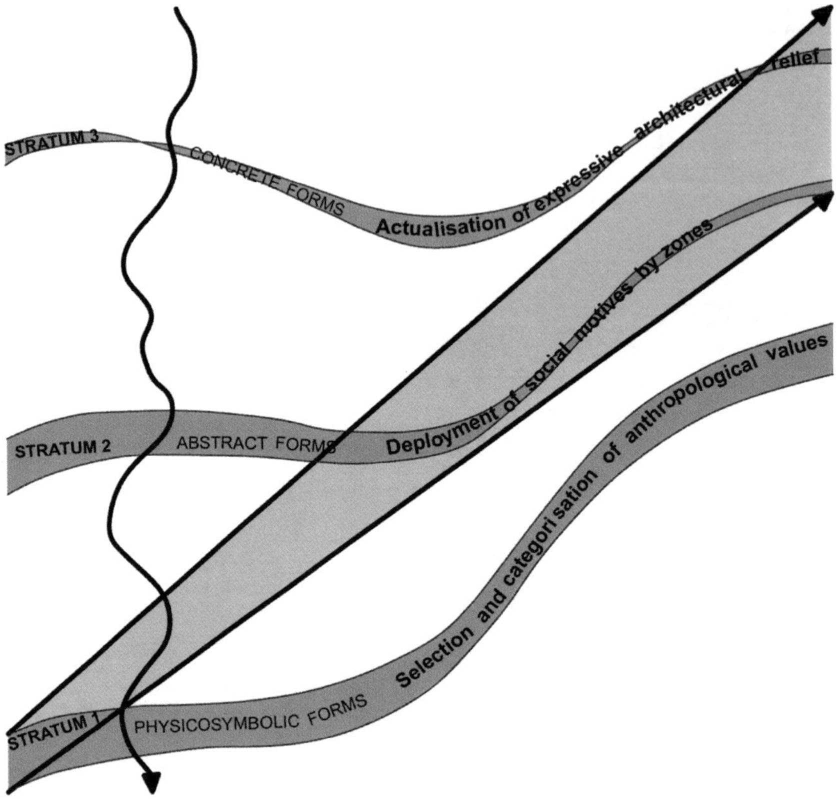

Figure 1. The city is a dynamic topology which is deployed in history: in this figure are disposed the three strata mentioned in the text – spatiotemporal stratification.

Stratum 1 – corresponds to a long-lasting physico-symbolic selection; the period of actualisation is spread over 500 years or more. In other words, the territory melts into a physical localisation of anthropological values. This base structure from then on allows the actualisation of symbolic values on the site's physical substratum;

Stratum 2 – corresponds to the deployment of physico-symbolic masses around two great thematic axes, respectively those of the order of Authority and those of the order of Production, the period of actualisation spreading over 100 years and more. During a long period of historical interactions, one can observe the arising of categorised paths which carry these values. These paths are crystallised in axes around which are built edifices of diverse orders, of a political and religious nature, of popular manufacture or of a commercial and mercantile character. This intermediary stratum is called “abstract forms”;

Stratum 3 – is the actualisation of forms which enables the understanding of the architecture and of the city as an identifiable whole of fields of matter composing it: spatial, sociocultural, physical, etc.; the period of actualisation is spread over 30 years or more. The study of the strata during the course of centuries allows a distinction of fundamental stages of a structural modification which is set off in the Middle Ages, stabilised during the Renaissance period, and finally consolidated during the Baroque period. In actual fact, this formation enabled the deployment of diverse social groups all along the established axes. The network deployed by the streets, the squares, the palaces, the buildings of religious and political order, underlines this dynamic structuring which finally converts itself into an aesthetic activity of intercultural representation.

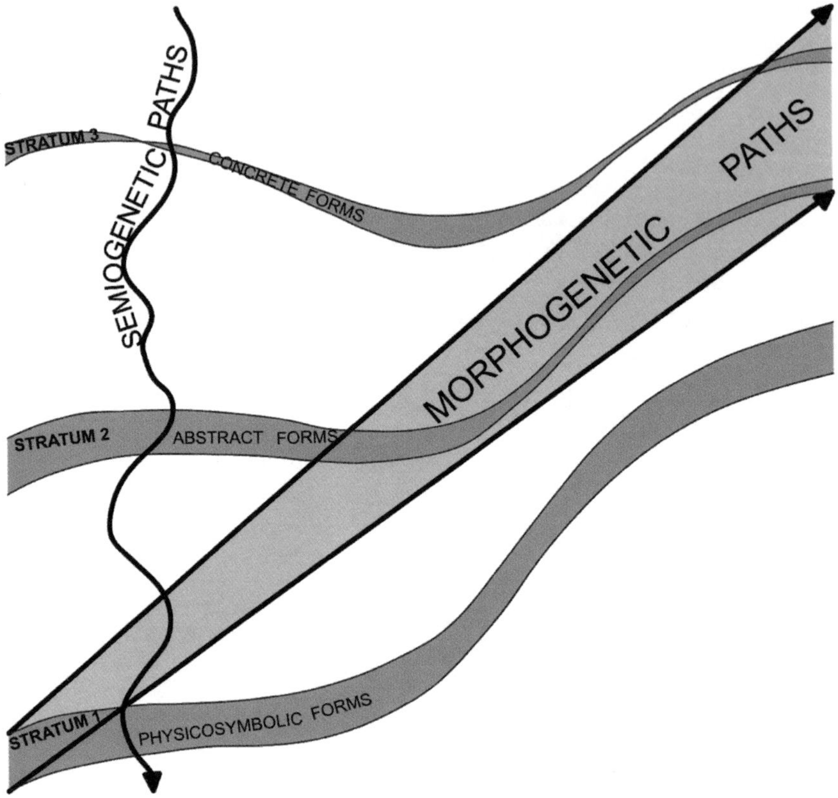

Figure 2. The stratification of three temporalities implies two movements: The morphogenetic and the semiogenetic paths.

2nd Inference

To think out Lisbon as a stratified depth which is elaborated according to three levels of temporalities and of a different nature, implies research on the process at work of structuring and reactualisation. Two movements can be distinguished (*Cf.* Figure 2):

- The first one, a stable movement: it is the one which engenders the city forms, that is to say the morphogenetic path[1];
- The second one, an unstable movement, is the one which engenders the city forms, that is to say the semiogenetic path.

3rd Inference

From these stratified paths emerges a schematisation which enables the display of how the diverse levels of topologisation of significations can be convoked simultaneously like the diagram – the *cusp* – demonstrates. This schematisation of the urban universals is syntactic according to their functions and semantic according to their constitution.

The section entitled *3rd Inference* is based on the study of figure 3, this being an *"iluminura"* surrounded by an architectural decoration of a Manuelin style (late Gothic), extracted from the *Crónica de Dom Afonso Henriques* written by *Henriques de Duarte Galvão* at the time of King *Dom Manuel*. The chapter is organised in two parts:

- In the first, t*he unfolding of the form of Lisbon*, the syntactic dimension of *Urban Universals* as diversity is presented. This diversity of strata is brought back by the *cusp*[2] by units taken care of by the semantic dimension.
- In the second part – *Urban Universals*, the diversity constructed by the reading of the image examined in the first part is linked to the unit aimed at by the semantic dimension of *Urban Universals*.

The notion of *Urban Universals* boils down to conceiving society as a dynamic topology of forces. This notion regroups simultaneously the spatial

1 The morphogenetic path is a central concept in semiotics. In presenting it as a process of engendering according to a triple stratification going from the most profound to the most superficial, the concept of a morphogenetic path of the human establishment comes closer to what is named as "generative path" [le parcours génératif] in A.J. GREIMAS's semiotic theory.

2 The cusp, also called the crease (cf. Figure 5 - The Urban Universals) possesses two curves which describe the behaviour of the two attractors (A, B) as spatiotemporal stratification. Such a stratified space of control enables the introduction of new functions, these new functional phenomena figure then like paths in the catastrophe. If we introduce the path going from the critical point towards C, the well is folded and is doubled up in (A, B) and passes towards the external side of the cusp (Brandt 1992: 72).

organisation and the social organisation. The hypothesis to be presented all along this chapter is the following: *between (1495–1521) the reign of Dom Manuel is simultaneously a project of monumentalisation, a project of society and a project of sovereignty.* The syntactic and semantic depth of the whole of these events must be demonstrated simultaneously. The *cusp* is the schematisation which is fitting for such a demonstration (*cf.* Figures 5 and 6 – *Urban Universals*). The reign of *Dom Manuel* marked a key moment in structuring a new style of society, that of the Catholic monarchy[3]. This syllogistic schematisation which is deployed in history is the object of the demonstration put forward (*Cf.* Figure 3).

2. First Part – The unfolding of the form of Lisbon.

Let us summarise the situation of the years 1495–1521. During this period, society managed to find a balance between the diverse forces in conflict and this thanks to an urban programme of monumentalizing. The reign of *Dom Manuel* marks a key moment in the process of structuring Lisbon. He transforms his project of monumentalizing into a project of forming society, which will allow him later to become the dominant actor. This seizing of power by the King was made possible thanks to a new system of positions within the space of various actors of society, which gave birth to quite notable urban transformations. The discovery of a maritime route to India undoubtedly lead to important social, economic and architectural upheavals, including the creation of a new style of architecture – the Manueline (Manuelino). The urban transformations will enable to balance out of resulting conflicts and the creation of a new system of stabilities. In 1521, Lisbon is no longer a medieval town. We are now faced with a new urban scenario and, consequently, a new concept of city and we shall thus enter the Epoch of the Renaissance.

The below *iluminura* has been chosen because it can explicate the proliferation of related events which will make the city form evolve towards a new system of stabilities, a form triggered off and embodied by the King.

---

3 Catholic monarchy: government of a sole King, catholic and universal, according to the meaning evoked by Serge Gruzinski; cf. notably his work (2004), Les Quatres Parties du Monde. Histoire d'une mondialisation. Éditions de la Martinière, Paris.

Figure 3. A panoramic view of the city of Lisbon in the 16th century. Source: "Crónica de Dom Afonso Henriques" of Duarte Galvão (Musée Condé de Castro Guimarães, Cascais).[4]

4 The reference to this illumination comes from a catalogue of an exhibition managed by: Moita, Irisalva, (1983). Lisboa Quinhentista. A imagem e a vida da cidade, CMI, Lisboa.

From an empirical analysis of Figure 3 it can be affirmed that there are several salient forms[5] which stand out from this illuminated picture. The most evident and most surprising form is that of the Crusaders at the time of the reconquest (from the Arabs) of Lisbon in 1147, whereas the image in question represents it much later, in the 16th century. The Crusaders come forth from there, introducing then onwards a spatiotemporal movement between these two historical moments (1147 and the 16th century). One can see them on either side of the King's Palace of *Alcaçova*, a very significant form of representation for the present analysis. If one follows this reasoning by a comparison of this map with the whole of the maps of Lisbon of this period, evident distortions can be seen which produce, on purpose, a deliberate and meaningful symmetry. This is why it is possible to postulate that this symmetric and pyramidal structure is the key to the reading of the image as it serves to underline a limited ensemble of salient forms, the Crusaders being a salience setting off the central theme.

In order to complete this postulate, it is necessary to specify: in actual fact there does exist a map[6], very similar[7], a quasi-copy according to certain aspects, which also represents two armies around Lisbon, but on which, surprisingly enough, the significant distortions evoked, shine by their absence.

What are the elements observed (*cf.* Figure 3) which produce the symmetric structure of the *iluminura*?

- The context in Manueline (*Manuelino*) style confines the space of the representation of the city. The arc at the top of the window opening out on the landscape and the two lateral columns on the page of the book underline the symmetry of the illumination.
- The symmetry is intensified by two clouds situated laterally in an ordinary light blue sky as well as by two military camps (Crusaders) disposed on each side on the two green hills. The city, centrally positioned, in a triangular form, is closed in by fortified walls. From all that, comes into view a central axis around which are disposed the principal actions of the story that the picture seems to want to tell us.

5 By definition, a form is said to be salient if it is detached from a background (Thom, 1991: 104.

6 Cf. Antonio de Holanda and Simão Bening "Vista de Lisboa" Genealogia dos Reis de Portugal.fl.8, illuminura em pergaminho, British Library, London.

7 The two pictures, iluminuras (illuminations), have a common origin, they come from chronicles (crónicas) where the genealogy of the Kings of Portugal are presented.

What are the salient forms that stand out from this symmetric structure?

- The space of the *iluminura* is arranged in terraces of four regions: above the sky (light blue); in the top central position, the city (the dominant colour is light grey even if in this region there is to be found a profusion of colours, green, ochre, red, white, blue and brown); in a lower central position, the port of the *Ribeira* (dark grey); and finally, at the bottom, a vast lower fluvial zone (dark blue). One can assert that the space of the tableau is horizontally structured, according to an inverse symmetry, by two backgrounds, one light the other dark.
- The summit of the central axis of the image coincides with the higher angle of the triangle which is the city, angle constituted by the Palace of *Alcaçova* (the King's palace). The two extremities of the base of the triangle in question represent two salient forms which are the commercial square of *Rossio* and the *Sé* Cathedral. This well established base is underlined by the port area of the *Ribeira* River, an interstitial zone between the river and the city. If one connects the salient forms with the symmetric structure, it can be asserted that the interstitial zone is not only a frontier separating the two light and dark backgrounds, but also the base of a triangle in the interior of which there is a profusion of colours.

But based on this enumeration of coloured backgrounds and salient forms disposed symmetrically, we observe that the only elements really detached from this structure are the Crusaders camped in the greenery, which justly gives them a start off position for the sought after dynamics.

What are the abstract forms (stratum 2) which can be located from the figurative dispositions of the concrete forms (stratum 3)? At this level of pertinence, let us link them to the values which they carry to bring out the effects of meaning.

As it is a question of visual expression, how can the colours and their topological disposition in the figure be interpreted?

Due to the profusion of colours in the representation of the city, this seems quite the centre of the "narration". It is an intensified dimension by the double presence of the Crusaders in the course of reconquering it, whereas they bear its colours highlighted by the contrast of the green of the hills where they camp. It is as if this space depends on another "narration" and on another period.

An attentive examination of the colours brings out the following logic: the sky, light blue; the earth, light grey; the boundary (the port of the

Ribeira), dark grey and the river, dark blue. This suggests two couples of action: sky/earth, in two light shades; port/river, in two dark shades.

One can distinguish two themes: the internal space of the city and the external space of movements of caravels towards the New World. The New World, which is represented by ships, the "wooden cities" (*Campanella*), seems to have an enormous influence on the reorganisation of the city, because the proportion of the lengthy public square of the *Ribeira*, disposed horizontally, is like a long boundary separating and linking the town to the river (it is precisely in this area that will be established the monumental axis of the town).

If we return now to the representation of the Crusaders at the time of the Reconquest (1147) disposed on both sides of the King's palace of *Alcaçova*, whereas this picture was conceived later during the reign of *Dom Manuel* (1495 to 1521), the question of timing is inevitably raised: why want to represent an event of a past reign (during the reign of the first King of Portugal *Dom Afonso Henriques*), whereas the *iluminura* (illumination picture) dates from the reign of *Dom Manuel*? For this, one must again compare this picture with the more conventional representations of the Reconquest. It can be observed that the camps are located on the hills: on one side the hill *São Francisco*, on the other that of *São Vicente* and in the middle that of *São Jorge,* where the royal palace of *Alcaçova* rises up. How can we link the originality of this spatial disposition of the historical event which was the *Reconquest*, with the artificiality of this symmetry, a *meaningful distortion*? It is useful to specify that in the 16th century, contrary to its picture where one can distinguish, like in the 12th century, just a camp, the hill *São Francisco* was densely built up. The buildings which rose up on the hill at the time that the illumination in question was made are represented lower down like an extension of the *Rossio*. The convent *São Vicente de Fora*, the cultural centre of the town, disappears from the hill of the same name, just to make place for another military camp symmetric to its predecessor, there again like in the 12th century. As for the hill of *Sant'Ana*, set back from the others, it has completely disappeared from the entirely remodelled countryside! It is thus that a new purely virtual physical morphology is created having, as the principal characteristic, the positioning of a massively central hill, sloping down towards the River, and flanked on either side by a plateau.

What is the articulation of salient forms of each part of the *iluminura* which can be proposed from the information invoked above, and how can we link them to the values?

1. Representation from the point of view of the Crusaders: the devices for the construction of the image seemed to point to the fact that in the interior of the city there still subsist Arabs to be evicted like at the time of the Reconquest. Now, it is during the reign of *Dom Manuel*, between 1496–1497, that the Jews and Arabs were expelled and their respective districts closed. This picture then gives its real meaning only through the comparison of these tragic events, acquiring thus the true dimension of a narrative process.
2. Representation from the point of view of the Palace of Alcaçova: it has already been specified that the symmetric and pyramidal structure is an inescapable key to the reading of the picture. It represents, as a central position, a Palace unceasingly besieged and the seat of all the kingdoms, from Phoenicians (1280 BC) to Arabs and Jews (Aluxbuna or Olissibona) 714 AD to 1147 (by way of Lusitanians 700 BC, the Greeks 638 BC, the Celtics 300 BC, the Ligures 700 BC, the Romans (Olisipo) 60 BC to 409 AD and the Barbarians, Visigoths, Alans, and Sueves (Kutia) 409 AD to 714 AD). As from the Reconquest in 1147 the Palace of Alcaçova becomes the royal palace in Lisbon. Following that, King Dom Manuel is to be the first to want to change this order of things, foreseeing his installation on the Ribeira, on the shores of the Tejo, by the construction of a new palace. This radical decision, a symbolic double rupture as compared with the past, can also be interpreted not only as deliberate political will to redefine the position of the King's relations with the other social actors, but also to definitely separate himself from the Arabs that he had just expelled, added to which it is necessary to specify that the former royal Palace of Alcaçova has an Arab name signifying "fortress". All this seems to well and truly evoke the theme of sovereignty.
3. Representation from the point of view of King Dom Manuel's desire for sovereignty: The themes of commerce (*Rossio*) and religion (*Sé*), seeing their remarkable emplacement in the *iluminura* (illumination), are approached within the framework of the symmetry of relationships with the King. The King is placed above in his royal palace situated at the summit of the highest of the hills with, on his right, towards the bottom the commercial Square of *Rossio*, and on his left, towards the bottom, *Sé* (Cathedral). Thus there the king reigns over money just as over faith. By the topological positions, his relation with religion is situated at the same level as that of commerce. A paradoxical symbolic rupture as during the whole length of

history the religion of this city is in competition with the sovereignty of the King (the management of the city is at times assured by a bishop).

4. Representation from the point of view of a horizontal form of the long emplacement of the port of the *Ribeira*: it arises like a frontier between the town and the river. As for the orthogonal axis which is the embarkation quay, in prolongation of the central vertical axis descending from the Palace, it intensifies not only the theme of great discoveries but also magnifies the new relation of domination of the King towards other social actors as if he were the sole and unique moneylender of these maritime adventures.

5. Representation from the point of view of the discovery ships: these "wooden cities" go out to sea suggesting economic, evangelical and cultural maritime routes. On one hand, the vertical axis (*cf.* Figure 4) of this picture puts into perspective a palace, seat of all kingdoms, and the town which has been transformed in the course of time by the successive coexistence of different cultures; and, on the other hand, the same axis evokes the production of values that are merchant as well as religious.

Figure 4. The thematic values and the structuring activities of "living together": to Legislate, to Nominate and to Produce.

Through what has just been demonstrated, several salient forms staged by topological positions in the picture have been made noticeable. The articulation made has enabled the establishment of values induced by these forms. At this moment at the analysis of the *iluminura* (illumination), the significant forms must be linked one to the other by a network of significations. The thematised "figures"[8] enable a display of this network of significations. Within this perspective we discover that the three structuring activities of "living together" are respectively *to Legislate*, *to Nominate* and *to Produce*, each one corresponding to a figurative path. The following "figures" are to be retained (*Cf.* Figure 4):

- *To Legislate* (A) – politics, religion, and royalty;
- *To Nominate* (B) – the military, the feudal lords, the nobles and the aristocrats (in this picture, this position is particularly marked by the Crusaders installed in their camps around Lisbon);
- *To Produce* (C) – the people, the merchants, the shopkeepers and the bankers (in this picture, religion poses as a producer of values through its concern to evangelise the New World and to receive Tithes, we shall come back to this point).

At the syntactic level, the figures vary during the course of time. They depend on "the values of themes" which are updated at each moment of the evolution of the human group (Portuguese). It must be said that the reign of *Dom Manuel* marks the passage between a Medieval Society (founded on the thematic value of the sacred) to a society of the Renaissance (founded on the thematic value of a collective destiny which takes form as and when merchandises arrive). Thus we can distinguish between two types of "thematic values" (these values coincide with the critical point of the *cusp*):

- *Sacred* – Medieval Society;
- *Collective destiny* – Renaissance Society.

During the course of this first part of the chapter *The deployment of the form of Lisbon*, the syntactic dimension of *Urban Universals* as a diversity has been evoked, setting out the abstract forms deploying the social patterns into zones (stratum 2). Now it is a question of having an interest more particu-

8 For example, the theme of the production of values is taken in hand by the "figure" of the people, the merchants, the shopkeepers and bankers (in this picture, religion poses itself as a producer of values by its concern to evangelise the New World and by the receipt of the Tithes Tax).

larly in semantic positions (stratum 3). In order to do this, the diversity constructed by the reading of the illumination is linked to the unit aimed at by the semantic dimension of urban universals, which will be studied in the second part of this chapter. The three structuring activities of "living together", already set out above, coincide in the picture with the localisation of the *salient forms* already described (*Cf.* Figure 4):

- *To legislate (A) – The Alcaçova Castle*
- *To nominate (B) – Two Crusader camps (feudal lords)*
- *To produce (C): Commerce – the production site of market values; Sé, the cathedral – the site where religious values are created.*

In the next paragraph, the *Urban Universals* shall explicate the moment of structural displacement when King *Dom Manuel* becomes the dominant actor.

3. Urban Universals

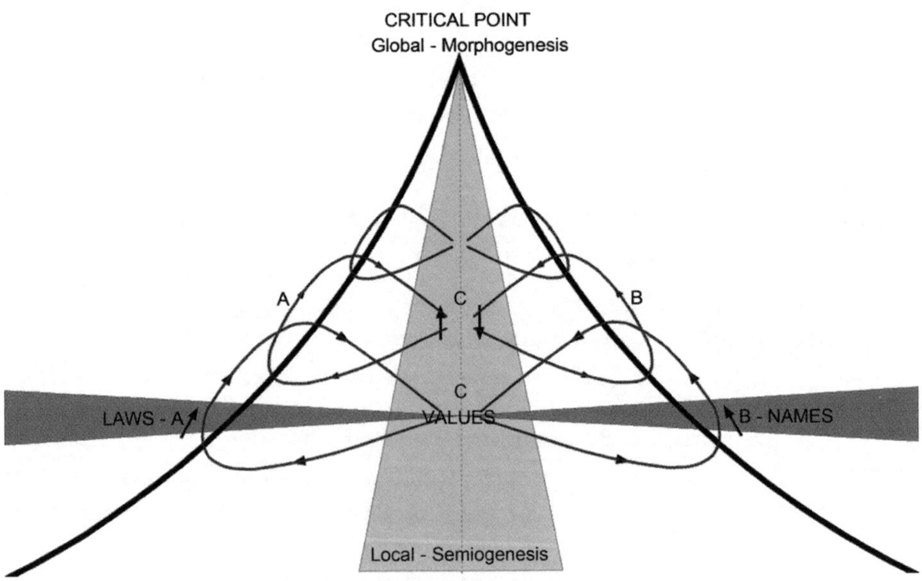

Figure 5. Urban Universals: Laws, Names, and Values.

The observation path of the *iluminura* (illumination) leads to the hypothesis maintained in this article according to which, *between1495–1521, the reign of Dom Manuel is a triple project of monumentalisation, of Society and of Sovereignty.* The syntactic and semantic depth of the whole of these events must be called upon simultaneously, and the *cusp* is a schematisation which fits well such a demonstration (*cf.* Figures 5 and 6 – Urban Universals); because, in actual fact, the *Urban Universals* are syntactic according to their functions and semantic according to their constitution. The notion of *Urban Universals* amounts to conceiving society as dynamic topological forces. This notion simultaneously regroups spatial organisation and social organisation. The *cusp* (*cf.* Figure 5) manipulates the topologisation of significations but does not produce these significations. *The semantic components which follow apply to any city: (A.) – Law; (B.) – Names; (C.) – Values.* Figure 5 disposes them topologically allowing the unveiling of the dynamics. It is only when the "figures" pass through the cusp that the *semantic components* are convoqued.

Remarks regarding the functioning of the cusp applied to the morphogenesis of Lisbon:

The *cusp*, also called the crease (*cf.* Figure 5 – *The Urban Universals*) possesses two branches which are like a spatiotemporal stratification. (A) – Law and (B) – Names are, from an analytical point of view, empty positions. Such a space of stratified control allows the introduction of new functions. These new functional phenomena figure then like paths in the catastrophe. In the analysis, the paths between (A) and (B) are described by the passage through the "figures". If we introduce the path going from the critical point towards C, the well is folded and is doubled up in (A, B) and passes towards the external side of the cusp.

In order to grasp well the new *catastrophic* mechanism utilised by the King Dom Manuel, it is necessary to well understand the relationship which is established between the three of the urban universals (A) – *Laws*, (B) – *Names* and (C) – *Values* on one hand, and the stratified paths brought up to date by each figurative path, to *Legislate* (politics, religion, and royalty), *Nominate* (military, feudal lords, nobles and aristocrats) and to *Produce* on the other hand (if in the illumination, the people, merchants, shopkeepers, bankers are producers of merchant values, on the other hand religion poses as a producer of other values through its concern to evangelise the New World and to perceive Tithes).

Through the observation of the behaviour of the urban universals (Marcos 1996), we can state that the focalised force (A) (*form of the Law*) is affected by its own tendency to be at rest. A Royal power coincides often with this position. The non-focalised force (B) (*substance of the Law*) operates either as a motor if the tendency of the force (A) is at rest, or as a barrier or a brake if it is in motion. For clarity's sake, let us call respectively, to begin with, these forces (A) and (B) *the form of Law* and *the substance of Law*. Brandt defines these two types of attractors as symbolic instances. Whereas (B) creates an "aesthetic" of the difference, (A) is linked to an "ethic" of equality. As a matter of fact:

The "feudal regime" creates the *difference* where (B) operates as a motor of society; the feudal lords protect the common people, defenceless from danger. (B) acts when the present conditions within a period are favourable to it. (B) awaits the present moment to propagate its movement. When this position is much stronger than that usually occupied by the King, it is characterised by the "breaking up of sovereignty into small pieces, which escapes from the central power and disperses itself between the hands of a certain number of great Lords"[9].

- The "absolutist regime" creates equality. (A) operates as a motor of society – when the King is too strong, he ends up by abolishing all possibility of *difference* between the other constitutive forces of society. The King "reigns over his court by marking out the distances, he holds in check the orders and the groups of his kingdom by watching and maintaining with care the balance of tensions[10]" (Elias 1976: 182).

(A) and (B) are the two versants of the "Authority". (A) is placed in relation to *royal blood* (acquired by birth), which gives it the legal competence *to judge* and *to argument* what is common property. It "uses the language's capacity to express itself under an impersonal form, and to formulate logical and legal conditions, to argument and expose general ideas": it is the "ethics" of equality. The King places himself at a "good distance" from the Court, this distance being always subject to renegotiation. (B) situates itself in relation to history based on the "blood of the family", that is to say blood shed in order to protect, until death, the common property of the society,

9 Le Petit Robert (1991: 770).

10 Le Roi "règne sur sa Cour en marquant les distances, il tient en échec les ordres et groupes de son royaume en surveillant et en maintenant avec soin l'équilibre des tensions" (Elias 1974: 182).

which gives it the competence to nominate the qualities of objects and subjects in creating *narratives*. As said by the historian *Borges Coelho*, royalty is founded on the blood of martyrs dying in order to enlarge the universe. Due to this fact, the "families of nobility" found a privileged place when they were gathered together around the Portuguese Court.

All belong to the same law, to a same community, which saves the Christian community.[11] This capacity of (B) to designate, to nominate, is then linked to (A) by the same basic law: *honour founded on Christian law*. As a hypothesis, we can put forward that a same law governs the honour of the conventions of knighthood. "The King felt attached by this law like any other knight[12]", even if, little by little, the "absolutist kings" transformed these conventions into kinds of courteous games[13]; (B) uses the narrative capacity of the language and its potentiality of constitution of proper names to designate and to thus enable to attract towards oneself the transmission of works and privileges throughout the generations, it is the "aesthetics" of the difference. Thus (A) and (B) are linked to a same law since they are the two versants of "Authority" (*cf.* Figure 5 – the horizontal axis in dark grey corresponds to trajectories described in the *cusp* when (A) and (B) enter into relations – it is the axis of authority).

"Commerce articulates the *difference* by (B) and *equality* by (A) in the contracts recorded in the accounts" (Brandt 2004). This articulation or "good distance" is possible thanks to the *market*, in the sense where (C), the merchants, are located at the threshold of the positions (A) and (B). (C), by its position, is the field of exchange of objects in quantity, "Productivity" (*cf.* Figure 5 – the vertical axis in light grey corresponds to trajectories described in the *cusp* between the critical points and the threshold represented by the attractor (C) – it is the axis of productivity).

The three positions (A), (B) and (C), deployed, are three symbolic positions which any urban organisation needs to characterise and distinguish itself. What in effect changes in time is the capacity to actualise these positions. This capacity is all the more important in that it allows the deployment of a specific space of communication throughout historic periods.

---

11 ["Mas se os homens não têm entre si precedências por natureza, têm-na por pertencerem ou não à comunidade que salva, cristã. Dentro dela, os Portugueses reunidos em torno da Casa da coroa de Portugal têm um lugar de eleição. Esta Casa funda-se sobre o sangue de mártires que a dilatam por todo o Universo" (Borges Coelho 1992: 135–136).]

12 Cf. Norbert Elias (1985: 158).

13 Cf. Norbert Elias (1985: 158).

The town would thus be an area of communication made current by the structural "base" of positions (A), (B) and (C). As already indicated, these positions maintain between them geometric relationships. The question is to know what is the "good distance", internal and external, for putting into place these positions; more precisely, what is the "good distance" in order that the dynamic capacity of these positions be effectively brought up to date?

The distance is woven by the rituals of circumstances. It is the emergence (C) situated on the "threshold" between (A) and (B) which regulates the shape of this distance: it becomes, by this fact, fundamental in the sense where it expresses the balance between two other positions (A) and (B). According to Figure 5, it appears that the "market" is situated at the crossing of the axis of "Authority" and the axis of "Productivity".

The long path of clarification of the functioning of cusp as regards to the morphogenesis of Lisbon, is imposed on the analysis. As demonstrated, the Urban Universals are fundamental for undertaking a semiotic analysis of the deep structures, 1st stratum. Nevertheless it is necessary to continue the semiotic analysis of the illumination in order to demonstrate, that which has only just been evoked, that the reign of Don Manuel is in effect well and truly a triple project of monumentalisation of society and of sovereignty. It is not a question of results provoked by the sole political will because this being has benefited from a whole lot of structural conditions which he knew how to exploit very ably. Let us look at how the profound structure reveals itself as and when the historic events link up with the significant forms disposed on the urban substrate.

The Great Discoveries during the reign of *Dom Manuel* begin to yield considerable sums of money. We live a very special historical moment concerning the balance between the diverse social actors. We observe that in this period, position (C) has definitely the upper hand on attractor (A). This historical moment makes the King a distributor of funds instead of a distributor of land. It is in the breach created by this change that the middle-class, by rushing in, enhances its position in society. It required time for the nobility (attractor B) to adapt itself to this difference. During centuries, they had a large autonomy in relation to the King; for him, it was much easier to suppress a pension in money than to take back a fief or ground rent. When we linked the historic events with the symmetric structure of the *iluminura* (illumination), we saw the appearance of a vertical axis around which were disposed salient forms of the order of production of values. The axis stages and places the King (represented in the picture by the Palace), then the city and afterwards the discoveries. The disposition of this axis of production of values underlines, on one hand, the importance of the position (C), and, on the other hand, shows clearly the alliance made by the King with position (C) to the detriment of nobility (B). From the point of view of the disposition of the picture, the nobility, represented by the Crusaders, seems to be outside of the principal theme, and even as if outside of the times. For what reason?

– One must be aware that one of the functions of the nobility was defence (represented by the Crusaders). Now, money allowed the King to hire mercenaries. This change in the economy weakened the position (B) of the nobility. The shopkeeper class had been greatly advantaged by the long lapse of time necessary for the restoration of their dominant position (B) in placing themselves at the court. At the time of the Renaissance, the possibility of reducing all the symbolic positions into one, merchandise, arises. According to *João de Barros* cited by *Borges Coelho*: "Make as many divisions as you wish as to the characteristics of men: whether they be ecclesiastics or laymen, whatever their rank, their nationality or their profession, none can live without buying or selling[14]". The author continues in his commentary by underlining the power that merchandise exerts over all the social groups. "No one can live without buying and selling, the truth is that someone must produce goods. Effectively, this task is attributed to the domain of producers (…). The principal action of this period of merchandising brings out the place of each one in society: the Priest prays, the Knight defends, the Lord governs[15]".

Let us observe once again on the *iluminura* that the Crusaders (nobility – B), positioned slightly below in relation to the King (the Palace of Alcaçova), recall to him, by their proximity, the blood shed in order to defend the kingdom. The strategy of the pictorial devices presents the nobility as an position on which the King depends in order to reign, since nobility represents the *substance of Law*, and their importance in the *iluminura* is perhaps the key for the reading of this representation displaced in time (it must not be forgotten that the dynasty of *Dom Manuel* was founded in 1385 by the *Cortes de*

14 ["Faze quantas divisões quizeres de todas as calidades de homens, quer sejam eclesiásticos quer seculares com quantas dignidades, estados e ofícios houver entre eles, nenhum vive sem comprar e vender" (Borges Coelho 1992: 78).].

15 ["Mas se ninguém vive sem comprar e vender, a verdade é que alguém tem de produzir as mercadorias Essa tarefa cabe às "colunas da república", [...]. Ca mediante o suor de todos estes, o sacerdote reza, o cavaleiro defende, o senhor governa" (Borges Coelho 1992: 79).].

Coimbra on one hand; and that during the election of King *Dom João I*, nobility had played a fundamental role on the other hand). Through the devices for the construction of the *iluminura*, we can deduce that the Crusaders at the time of the Reconquest are represented there to indicate the presence of the Arabs within the interior of the city, people who must from then on be expelled by Royal decision. Because it is during the reign of *Dom Manuel*, between 1496 and 1497, that Jews and Arabs are expelled by the closing down of their respective quarters. This question of expulsion of Jews and Arabs is intrinsically linked to the hypothesis of "repatriation on earth of an economy of salvation"[16]. Let us look how:

By this repatriation, the Christian Portuguese shopkeepers can from then onward achieve their salvation not only in the afterlife but also here on earth. The reasoning can be summed up in the following way:

- On one hand, the church, substituting itself for the Jews, allows usury loans to Christians (which they forbade previously) but, on the other hand, usury is a serious fault which contravenes its precepts. It is necessary, therefore, to make up for this fault. The reparation consists in bequeathing to the church by testament the money illicitly earned following the death of the person. Thanks to this money, the church achieves its programme of urban monumentalisation in parallel to the domain deployed by the King in front of the river, a space represented on the *iluminura* by the port of the *Ribeira*. This society was "still very dependent on the clergy, "the doctors of the soul". (…) By the indirect means of the annual commemoration of the deceased, certain possessions become the property of the Church. Besides it is a major preoccupation of "the doctors of the soul"[17]". "The collectors of ecclesiastic tithes provoke terror in the people. The simple origin of the verb 'to decimate' which signifies

16 The hypothesis of "repatriation on earth of an economy of salvation", introduced by Gilles Ritchot (1991) and Gaëtan Desmarais (1995).

17 "O clero, os médicos da alma. [...] Pelas comemorações anuais dos defuntos entram terras e bens para a Igreja, o mesmo acontecendo com os trintanários do Santo Amador. O "grosso", eis a preocupação dos médicos da alma. Já sabemos que "por bancos de câmbio" é que "posso saltar no curral das mitras". E os padres "andam alugados, louvando o santo de quem se faz a festa"" (Borges Coelho 1992: 79).

'to eliminate' illustrates well of the attachment of the Church to money at this period[18]".

The "repatriation on earth of an economy of salvation" supposes an alliance between the King and the Bishop, hence the localisation of the new Royal Palace (1521) in front of the river, linked by the commercial high-street (1294) to the cathedral (1150) on the hill *São Jorge*. Another consequence of this repatriation is the passage to commercial capitalism, with trade towards the New World and the Great Discoveries. The success of the voyage of *Vasco da Gama* (1495) towards the Orient and the resulting economic policy makes Lisbon a warehouse for precious goods[19]. The Great Discoveries provoke an afflux of goods and intensify at the same time conflicts already existing between the diverse religious confessions (Christian, Moslem and Judaic).

This economic programme of salvation conditions the spatial extension and the monumentalisation of "the axis of Authority", axis which runs along the riverbank. We note that this axis composed of political and religious domains deployed from East to West and parallel to each other; in their interstices, the Middle-Class (C) comes to settle down. This axis corresponds to a continued key of spaces edified by those who represent *Law* (A) and *Names* (B). On the *cusp* below (*cf.* Figure 6) this axis corresponds to the horizontal axis in dark grey.

18 "Os dizimeiros da dízima eclesiástica. Razões tinha no nosso povo em carregar de terror o verbo dizimar que à letra não é mais que achar a décima parte ou dízimo. "A hipocrisia está com a cabeça cheia de mitras" [...]" (Borges Coelho 1992: 80).

19 ["O bom êxito da viagem de Vasco da Gama e a política económica dele consequente fizeram de Lisboa o entreposto das preciosas mercadorias orientais e ultramarinas" (Matos Sequeira 1947: 327).].

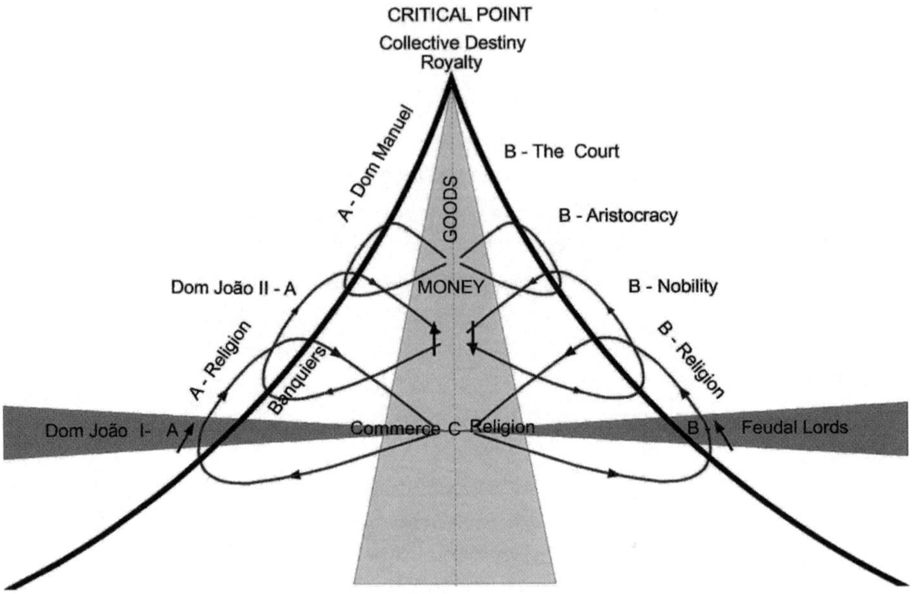

Figure 6. Urban Universals: Laws – politics, religion, loyalty. Names – the military, feudal lords, nobility, aristocrats. Values – the people, merchants, tradesmen, bankers.

One must retain from this topological setting the idea of considerable development of "the axis of Authority" (East-West). This axis is represented on the illumination by this long borderline space on the riverbank (in dark grey). This axis is organised by the social actors who have established themselves there, attracting there even the "assembly" of religious communities and the upper classes ("Political and Religious"). This development was based on the programme of an economy of salvation repatriated on earth.

Parallel to the development of the "axis of Authority", there is the development of "the axis of Productivity", that is to say the "concentration" of "the people" towards the North (*cf.* Figure 6, this axis corresponds to the vertical axis in light grey on the *cusp*). The crossing of the two axes configures an emerging morphological structure. This structure is engendered by the internal dynamic topology to the urban morphogenesis. It is not conceived and planned from the exterior by a social actor, but it results from a spatiotemporal stratification of a global process. This global process corresponds to the morphogenetic and semiogenetic path of Lisbon (cf. Figure 2 and observed on Figure 5 the positions of the morphogenesis and semiogenesis in the stratified schematisation – the *cusp*).

Valorisation and edification are major morphogenetic processes during these periods. The abstract morphological structure is in place since the end of the XVI century. Afterwards, the setting up of concrete shapes operates in accordance with positional values invested in the abstract forms.

In terms of topological displacements operated during the reign of *Dom João II* and consolidated later by the monumental buildings of *Dom Manuel* (1495 to 1521), hypothesis according to which the King has become the dominant social actor of society through his triple project of monumentalisation, of society and of sovereignty, is otherwise demonstrated or at least robustly established.

This seizing of power was achieved by virtue of the installation of a new system of spatial positions between the social actors. The "thematic value" of the Discoveries and the increase in commercial revenues which they procured would have rendered possible the "figurative" displacements. It is finally at this period that the two axes take a more consolidated concrete form (*cf.* Figure 6 – it should be observed on the *cusp* how *Dom Manuel* gets nearer to the critical point and entertains other types of relationships with the diverse social actors).

The internal process[20] is expressed by a path in the external space of the *cusp*. The axes of "the Authority" and of "the Productivity" would be expressions of these dynamic paths. They organise the figures around the two thematic roles on the external space, and they express the displacements of the "figures" in the internal space. The internal state is selected by the process of actualisation (the semiogenesis). This process coincides with the characteristics of an epoch.

The epoch is marked by the dynamic conflict present in the Renaissance society. The Manueline style is the expression of this dynamic. This conflict expresses the ambiguity itself of the style by underlining the formation under way of a new visual culture:

- a *tradition*, the classic orientation of late Gothic taste;
- an *innovation*, the constitution of a hyper-realism of visually organic tastes.

---

20 We can conceive these "figures" as being external states globally defined by an internal process. These states or "figures" are thus defined reciprocally by relationships of competition.

Several "approaches" have been sketched out to give an account of this dynamic conflict by means of schematisation that the *cusp* authorises. If the same concept of schematisation is applied to the forms of the human establishment, the edification (as concrete forms) is the result of a process morphogenetically linked to *social designs* (as forces), of which the dynamism would explain its consecutive constraints, taking root even more in the city's *physical substratum*.

In observing this emerging morphogenetic structure, we become aware of the power of the geo-historic path of a city, in a sense where this structure depends not only on the actants that valorise it, but also on the structuring itself, that is to say on the dynamic metabolism internal to its form. The structuring process of a long historical period in Lisbon (Marcos 1996) enables the understanding of how certain concrete forms can upset or to the contrary maintain the emerging structure (the analysis proposed in this article only covers a particular period). This double positional convocation we call the internal dynamic between concrete forms and abstract forms.

This gathering together assembles the elements which enable the representation of the form of the city to be conceived through the cartography of the three evolving strata. It enables the schematisation to be approached as morphology of the open universals. We have shown that the topologisation of the significations arrives at the schema only by passing through the "figures". In fact, the "figures" are geometrical positions crossed by values carried by the "figures" during their paths.

From the potential stratum – the central organiser – unstable units are deployed that engender the morphology of the *polis* (*cf.* Figure 6). If we place ourselves on the horizontal axis of the cusp, the human establishment will be stabilised by the chain of categories as choice of position on the conditions of the physical substratum, whereas if we situate ourselves on the vertical axis, this corresponds to a process of engendering forms of the human establishment.

References

Borges Coelho, António (1992). *Tudo é mercadoria, sobre o percurso e a obra de João de Barros,* Caminho, Lisboa.

Brandt, Per Aage (1992). *La charpente modale du sens. Pour une semio-linguistique morphogénétique et dynamique,* Aarhus University Press, Aarhus.

Brandt, Per Aage (2004). *Spaces, Domains, and Meaning. Essays in Cognitive Semiotics,* n°4, European Semiotics, Peter Lang, Bern.

Desmarais, Gaëtan (1995). *La morphogenèse de Paris des origines à la révolution. Une analyse morphologique, dynamique et sémiotique,* l'Harmattan, Paris.

Elias, Norbert (1976). *La dynamique de l'occident,* Calmann-Lévy, Paris.

Elias, Norbert (1985). *La société de cour,* Flammarion, Paris.

Gruzinski, Serge (2004). *Les Quatres Parties du Monde. Histoire d'une mondialisation.* Éditions de la Martinière, Paris.

Holanda, António and Bening, Simão "Vista de Lisboa", in *Genealogia dos Reis de Portugal,* fl.8, iluminura em pergaminho, British Library, London.

Marcos, Isabel (1995). "Semantic Density, Topological Density. A view of sixteenth century Lisbon", Semiotik, in: O. Møystad, (ed.), *Nordic Journal of Architectural Research*, 4, Oslo: 51–64.

Marcos, Isabel (1996). *Le Sens Urbain. La Morphogenèse et la Sémiogenèse de Lisbonne. Une analyse catastrophiste urbaine,* doctoral thesis at the University of Aarhus in Semiotics of architecture (directed by Prof. Per Aage Brandt, Aarhus.

Marcos, Isabel (1998). "Le trajet de pèlerinage au couvent de S. José de Ribamar", in: *Sémiotique & Bible*, 90, Lyon: 43–62.

Marcos, Isabel (2002). "Macao Transocéanique. Quelques réflexions sur le découpage", Actes du Congrès de l'Association Française de Sémiotique: Sémio 2001, in: Jacques Fontanille (ed.) *Des théories aux problématiques*, Pulim, Limoges: 197–201.

Marcos, Isabel (2005). "Resignificar a forma da cidade: Do percurso ao conceito de património", in: Márcio Moraes Valença e Maria Vaz da Costa (ed.) *Espaço, Cultura e Representação*, EDUFRN editora da UFRN, Natal: 169–180.

Ritchot, Gilles (1991). "Études de géographie structurale", in: *Cahier Spécial,* 15, Centre de Recherches en Aménagement et en Développement, Université Laval, Québec.

Robert, Paul (1991). *Dictionnaire. Le Petit Robert,* Dictionnaires Le Robert, Paris.

Thom, René (1991). *Predire n'est pas expliquer,* Esthel, Paris.

Catastrophes: What are we talking about?

ÁNGEL LÓPEZ-GARCÍA

The *catastrophe theory* (CT) developed by René Thom beginning in 1972 is the only mathematical model that has proved adequate to natural sciences and to human sciences at the same time. One might expect such a model to be highly successful given that unified consideration of empirical facts is an essential desideratum of science. Nevertheless, things have not turned out as expected. Natural scientists tend to consider CT as a qualitative model, hence, as a model that does not allow one to make predictions on the structure of reality. By contrast, human scientists have frequently adopted it as an alternative formalization of semantic facts, together with much other formalization, passing over its mathematical complexity and using it in rather a clumsy way. Consequently, Thom's model continues to be somewhat peripheral to science even though there are several enthusiastic scholars that have done well within this paradigm.

Mathematicians are rarely interested in natural language. Sometimes they consider the relationship between Mathematics and Logics, as Russell and Whitehead did, but natural language is too vague, metaphorical and ambiguous for formalization purposes. Except for specifically formal grammars, which content themselves with the formal aspects of syntax, the remaining current grammatical theories have evolved from a more formal approach to a less formal one. It seems as if linguists had repeatedly discovered that formalism does not work in linguistics and that natural scientific methods cannot be transferred to the domain of language. This was the case with generative grammar when comparing the *Syntactic Structures* –1957– or the *Aspects of the Theory of Syntax* –1965– models with the *Minimalist Program* –1995– and was also the case with structuralism when confronting Hjelmslev's *Prolegomena to a theory of language* –1943– with Jakobson's *Essais de linguistique générale* –1963– or Halliday's *Language as a social semiotic* –1978.

On the contrary, Thom proposed a mathematical model which has internal phenomenological power, thus it is a model that relies on linguistic roots because language is the all embracing structure that gives form to our perception of the world (Thom 1983, 7). Natural sciences deal with facts, but human sciences deal with phenomena. A phenomenon is a fact that has been perceived, and is later stored in the mind and, hence, is known. Natural language describes phenomena: *le chat mange le souris* (*the cat is eating the mouse*), a favourite example of Thom, does not represent a fact, but rather the perception of a fact, which is a phenomenon. Do we need two different sets of mathematical notions, one capable of formalizing facts and the other for dealing with phenomena? Thom thinks that catastrophe theory underlies both epistemological domains. If so, criticism of CT from the viewpoint of natural scientists would be misplaced, but ignorance of CT by social scientists would constitute a pitfall.

As Thom pointed out (Thom 1983: 7; 1992: 13), any scene we perceive is structured by enhancing a salient element detached from the remaining ones that constitute the ground. Language and perception constitute two sides of the same problem because language processes are structured following the laws of perception. Actually, there is a large tradition of linguistic studies that are organized in terms of Gestalt Psychology from some Praguean scholars, such as Bühler (as a forerunner) and Jakobson, to generativists like Jackendoff or cognitivists like Langacker. This parallelism is encouraging and opens the way to considering linguistic structures as living patterns rather than merely random forms that follow each other in a chain.

However, Thom (Thom 1983, 37) also affirms that, despite its proximity to perception, language guarantees the irreversibility of the phenomena it describes for it freezes the state of affairs it refers to, something that perceptual images – for example the visual ones – cannot. In other words: Thom has put language over any other representational procedure as if language were not only a necessary condition permitting humans to grasp reality, but also as a sufficient condition for the emergence of science.

This is one of the most misleading aspects of his theory. Thom says that irreversibility is a consequence of language and, moreover, that irreversibility is in the grammar. But returning to his favourite example, *the cat is eating the mouse*, what should be considered irreversible, either the process of a mouse being eaten by a cat in the external world, which it certainly is, or the utterance *the cat is eating the mouse*? To my knowledge this utterance, as any other utterance, is reversible in the speech act: the speaker begins by a meaning he/she aims to communicate to the hearer and ends by uttering a phonetic chain relative to it, in this case *the cat is eating the mouse*; conversely, the hearer begins by recording the above chain and ends by recovering its meaning. No linguistic phenomenon would occur should the reversibility condition of language use fail.

Brandt (Brandt 1984, 60) helped to solve the above contradiction in CT by distinguishing between speech (*parole*) and language (*langue*). When considering utterances, speech is irreversible because a sentence that is uttered always modifies the mind of the hearer, whereas when we consider language facing words, where sound and meaning correlate, it is reversible. Similarly, Linguistics behaves like Physics: it deals with reversible entities at the microscopic level, but with irreversible ones at the macroscopic level. Thom agrees with Brandt but he makes moreover a very strong assumption: that the irreversibility of physical phenomena at the macroscopic level would be a consequence of the irreversibility of their linguistic expression (Thom 1972, 23–24).

Accordingly he suggests that elementary catastrophes constitute an ontology which underlies some mental processes that are manifested by linguistic expressions. Wildgen (1999, 192) pointed out that this remark implies that mental process comes first and linguistic expression comes after. Then he considers that semantic roles constitute a cognitive classification of those mental processes that are verbalized. Catastrophe theory enables us to obtain abstract images of these scenes, and by choosing a definite interpretation of the values of the parameters some localistic models are obtained.

To start with these abstract schemes (the fold, the cusp, the swallowtail, the butterfly and the three umbilics) and to obtain variations by changing parameters, as Wildgen (1982) did, has been one of the most impressive achievements of the theory of catastrophes. Nevertheless one could think that the seven elementary catastrophes accommodate language as a whole, which is from a descriptive point of view, but not in their dynamic development. For example, we feel intuitively that the schemes for *to begin* and for *to finish*, which are based on the fold, look acceptable, and also the schemes for *to capture* and for *to emit*, which are supposed to rely on the cusp, as shown in figure 1:

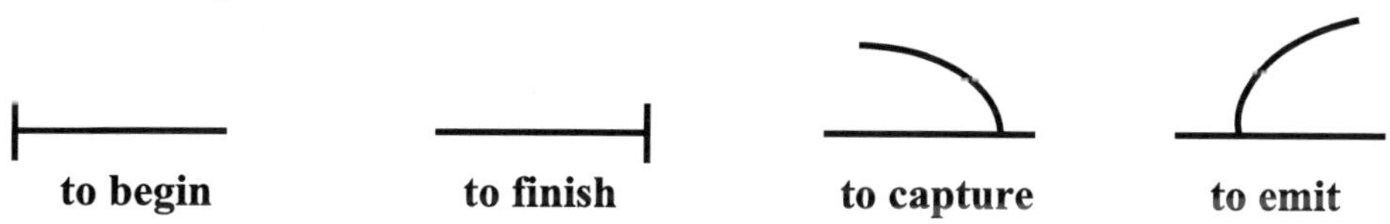

Figure 1. Four schemes for verbs.

Even the verb *to give* would be cognitively described quite well by the butterfly according to figure 2:

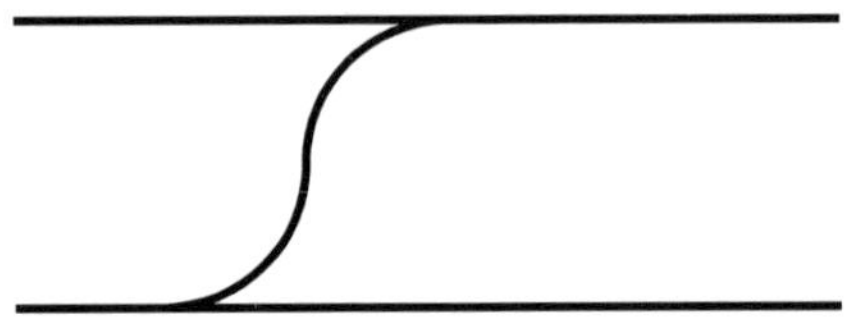

to give

Figure 2. The scheme for *give.*

but one wonders why *to cut off* should be described by the parabolic umbilic as it is in figure 3:

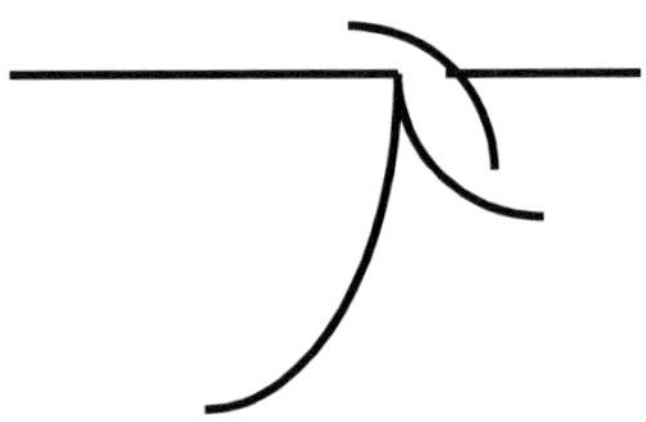

to cut off

Figure 3. The scheme for *cut off.*

Since *John cut the dress off* is a simpler sentence than *John gave Mary a book* and it is as complex as *Mary captured the dog*. I have the impression that the theory of catastrophes describes the real world rather than its linguistic mapping. And this impression is reinforced by the fact that scholars use the same schemes for these action verbs in several languages (they write *to finish/finir, to begin/commencer, to capture/capturer, to emit/emettre, to give/donner* or to *cut off/couper*, Wildgen, 1982, 32, according to Thom, 1977), despite knowing their translation is sometimes neither direct nor easy.

Petitot (1985a, 190) distinguishes three levels of modelling: the formal *catastrophe*, the *global scheme* where the actants reciprocally correlate, and, finally, its reading in terms of a *local* interaction of actants. The second one corresponds to Tesnière's stemma, a graph which exhibits the organic nature of syntactic nections. This would enable us to conclude that archetypes are preverbal schemes common to many biological structures. No compositional procedure underlies the combinations of actants that define each actantial scheme, they are rather irreducible gestalts. These biological foundations of the theory of catastrophes, applied to linguistics, constitute the very character that defincs it.

However, in relation to language, "biological" means "neurological", not only because the brain works by means of a set of neural webs, but also because the entire body receives nervous shots and hormones coming from the brain. Zeeman (1977) thinks that Thom's theory of catastrophes explains the mental activity that underlies language. Unfortunately it is not easy to attempt to connect modern theories of neuronal activity to current linguistic theories. The task of building a theory of the brain's language mechanism requires spelling out possible language mechanisms in terms of neuronal circuitry. The question arises now whether catastrophe theory provides us with such a model. Language and cognitive processes have been modelled on the basis of artificial neuron-like devices. It is obvious that such attempts conceive of language (or cognition) as a kind of assembly of discrete units that gives rise to more complex structures at higher levels. A catastrophe theory approach essentially differs from this view since forms do not exist before the constitution of the dynamic forces that tie them together (Thom 1990, 91).

Thom says that life is a pregnance and, for instance, hunger is a subordinated pregnance too. However, the linguistic display of a physiological feeling which is common to all human beings, hunger, differs greatly from one language to other. In the Spanish sentence *tengo hambre*, hunger is like an object possessed by the salient form. In the English sentence *I am hungry,* the salient form is in a state that consists of the pregnance hunger. In the classical French sentence* *j'ai de la faim*, the salient form participates in the pregnance. In Russian *ya, golodien,* both lexemes, the salient and the pregnant one, are simply put together. In German *mich hungert*, the concept of hunger is manifested as a verb, that is, as a catastrophe line where an attractor comes into conflict. Natural languages differ from each other in many ways, even when referring to a universal biological feeling like hunger: why should we suppose that the background of all these linguistic patterns is of a biological nature?

Brandt (1984, 82–95) faces up this challenge by comparing three kinds of translation, intralingual (from speaker's utterance to hearer's understanding), interlingual (from language A to language B), and intersemiotic

* Molière, *Le cocu imaginaire*, scéne IX: "Et moi, j'ai de la faim et de l'inquiétude".

(from code x to code y). All of them convey a meaning of meaning, they grasp things in a corresponding meaningful surrounding, and they are, then, of a topological nature.

I agree, absolutely. The problem is that reducing the meaning of meanings to worldly things reminds me of some semiotic attempts in the middle ages. The modists already distinguished among *modi essendi* – the real world – , *modi intelligendi* – the way they are conceived of by the mind – and *modi significandi* – their linguistic expression. A solution could be, as mentioned, to consider each modus as a kind of topology. A topology is a family T of sets which satisfies the two conditions: the intersection of any two members of T is a member of T, and the union of the members of each subfamily of T is a member of T. The set $X=\{U: U\in T\}$ is necessarily a member of T because T is a subfamily of itself, and every member of T is a subset of X. The set X is called the space of the topology T and T is a topology for X. The pair (X, T) is a topological space. Several topologies may be introduced in a set. If T and T' are topologies for X, then T is coarser than T' and T' is finer than T iff $T'\supset T$. Following this formal model, we could say that the set P of phenomena that are perceived by human beings can be measured by three topologies, the topology W of the things they perceive, the set K of the corresponding meanings they know, and the set L of the signs that describe them in a natural language, where L is finer than K, and K is finer than W. Unfortunately, such a proposal explains the methodological aspects of the problem, but it leaves the dynamic construction of meanings without a satisfactory explanation, for example, how are saliences and pregnancies (Thom 1988, 17 31) built up in whichever domain.

There is an alternative way to define a topological space, which is based on the notion of neighbourhood. A set U in a topological space (X, T) is a neighbourhood of a point x iff U contains an open set to which x belongs. Let U_X be the family of all neighbourhoods of x. Then: If $U\in U_X$, then $x\in U$; if U and V are members of U_X, then $U\cap V\in U_X$; If $U\in U_X$, and $V\supset U$, then $V\in U_X$; If $U\in U_X$, then there is a member V of U_X such that $U\supset V$, and $V\in U_Y$ for each y in V. Hence, we can define the concept of saliency in terms of a point (that someone has chosen) and its neighbourhood, which constitutes a pregnance: this interpretation of the concept of topological space would be analogous to the set definition mentioned above. This solution does not challenge the thomian position. Actually, Thom (Thom, 1990, 392–393) had pointed out that elementary catastrophe theory can be considered as an attempt to classify all possible types of analogic situations.

There are, however, two senses of the word *analogy*. Analogy serves as a means of explaining, to bring closer to the way of understanding. Analogy also has a meaning of likeness, identity, and proportionality. Analogy is, then, both a process and a structure. As a process it is a situation with an explicit non-understanding of something which has to be clarified and explained. As a structure it simply states a parallelism a/b=c/d, "a is to b as c is to d". The dynamic situation in explanation based on analogy has an implicit polarity which does not hold from a structural viewpoint: it explains learning vs. persons, known vs. unknown relationships. When examining the outcomes of the theory of catastrophes as applied to language, it turns out that the proposals that benefited from the dynamic aspect of the analogy above have generally succeeded. This was the case with the classical treatment of some linguistic developments: the dynamics of colour perception and the lexicon of colours (Wildgen, 1981), the dynamics of promise (Brandt, 1991), the dynamics of phonetic perception (Petitot, 1985b), or even the dynamics of conversation that I developed following the cusp model (López–García, 1991). In all these evolutionary trends a continuum of lexemes, pragmemes, phonemes or sentences occupies a regional unfolding with some catastrophic points, and the units change their position over time. Nonetheless, semantic archetypes do not change themselves into the neighbouring ones; they are not points in an unfolding. On the contrary, semantic archetypes are supposed to express the cognitive operations underlying language. Compare the use of the cusp model in both cases:

Figure 4. Comparison between “dynamic analogy” and “ static analogy”.

The image on the left represents a process A→ B where A has changed into B: for example, A would be a dialectal pronunciation of a word which changes to pronunciation B in the neighbouring region or in the future; it could also be a type of turn which evolves to another kind of turn during conversation. In the dynamic analogy, A and B never coexist, they follow each other. On the contrary, the image on the right represents a scheme where A and B coexist in a conceptual pattern: A would be the agent and B would be the patient of *Mary hurt John*, but *Mary* never changes into *John*. I wonder how we could employ the theory of catastrophes for both situations at the same time no matter if its topological definition fits all of them.

It is required that both analogical patterns, the structural and the dynamic, be simultaneously born at the origin of language. Wildgen (2007) proposed two principles: 1) The grasp schema constitutes a structural (bivalent) action schema with a long evolutionary history and includes a manner specification categorized on topological cues; and 2) The topologic-dynamical schema of grasping assembles causal/enabling/intentional meaning components, which are necessarily present in the purposeful shaping of a tool, and it also lays the groundwork for force-dynamics in phonic language. Let me add that these processes probably explain a first step, protolanguage, but are not enough to explain the full language step.

Any way, there is still a problem with the neurological layout of the catastrophe theory as developed by Thom (Thom 1983, 154–155). He said that this model describes the mental processes underlying speech acts. However languages differ greatly from each other, so I think catastrophes should describe rather the cognition of actantial schemes (which is presumably universal) than the cognition of their verbal manifestation. I find it difficult to maintain that the active and the passive represent opposite mappings of reality rather than opposite mappings of its cognition. For example, *Mary is taking the suitcase* and *the suitcase is being taken by Mary* represent the same process in the world, which could be represented by the cusp, and also the same cognition of such a process because one cannot conceive it in another way. Language relativity, either intralinguistic (Active. vs. Passive) or interlinguistic (Nominative vs. Ergative, for example) does not affect our thinking the scenes over but our way of expressing them. This means that the cognition of a scene can be more complex than its verbal representation, thus cutting something off could be thought following the scheme of the butterfly, although *he cut the tree off* follows the cusp. There are several categorizations of a particular cognitive experience depending on the linguistic structures that are employed by the speakers and /or the natural languages they have chosen.

The theory of neuronal group selection (TNGS) by Gerald Edelman (1987, 1988) constitutes an attempt to provide a neurologically plausible mechanism for alternative categorizations. The relevant fact of this theory is that associations are not embodied in specific combinations of neurons. On the contrary, the vast amount of neuronal networks are functionally degenerate, so any single function can be carried out by several non-identical neuronal configurations. This model enables us to bridge the gap between cognition and verbalization. While the genetic code provides for the anatomical topography of the brain, and stimuli overlap, linguistic wiring pathways presumably reorder these groups according to every particular verbal pattern. Instead of identifying the cognitive and the linguistic scene, the TNGS allows us to figure out the speech activity as a dynamic global activity, the so-called recursive synthesis, where the previously abstracted information, the concept, continuously receives re-entering linguistic maps, forming a global mapping.

Such a twofold analysis of cognition is by no means metaphorical. Actually, it is deeply grounded in neurological tradition. Hebb (1949) firstly proposed it when distinguishing *ignition* from *reverberation.* According to Hebb, a repeated neuronal activity produces durable changes that have the effect of dramatically increasing the facility to start action potentials. The activation of a neuronal web is called *ignition.* But after activation, the web does not completely vanish: the strongest connections yet allow a second wave of activity: *reverberation.* Reverberation is secondary to ignition from two points of view, temporarily and cognitively. As stated by Pulvermüller (2002, 169):

> Ignition is a brief event, whereas reverberation is a continuous process lasting for several seconds or longer. Ignition of a functional web involves all of its neurons, or at least a substantial proportion of them, whereas reverberation can be maintained by small neuron subgroups within the set being active at given points of time. Ignition does not imply a fixed spatiotemporal order of neuronal activity. In contrast, reverberation is characterized by a fixed sequence of neuron activations, a defined spatiotemporal pattern of activity within functional webs.

Fuster (1995) has extended the opposition between ignition and reverberation to the domain of memory: behaviour planning, which is located at the prefrontal cortex, has two functions: a retrospective function for sensory traces and a prospective function for motor programming. At this level both are short term memories. However such an approach – i.e the unified treatment of perception and action – enables us to extend neurological considerations to language. Linguistics is used to treating motor and perceptual explanations of the origin of language at the same time (Jackendoff 1987). Similarly, there would be some perceptual neuronal connections that fire globally – ignition – and some sequential connections that fire in a rather transitional way. Gestalt theory characterized them as two levels of representation, the figure-ground level (a global design where the figure stands out) and the boundary level (a transitional design):

ignition

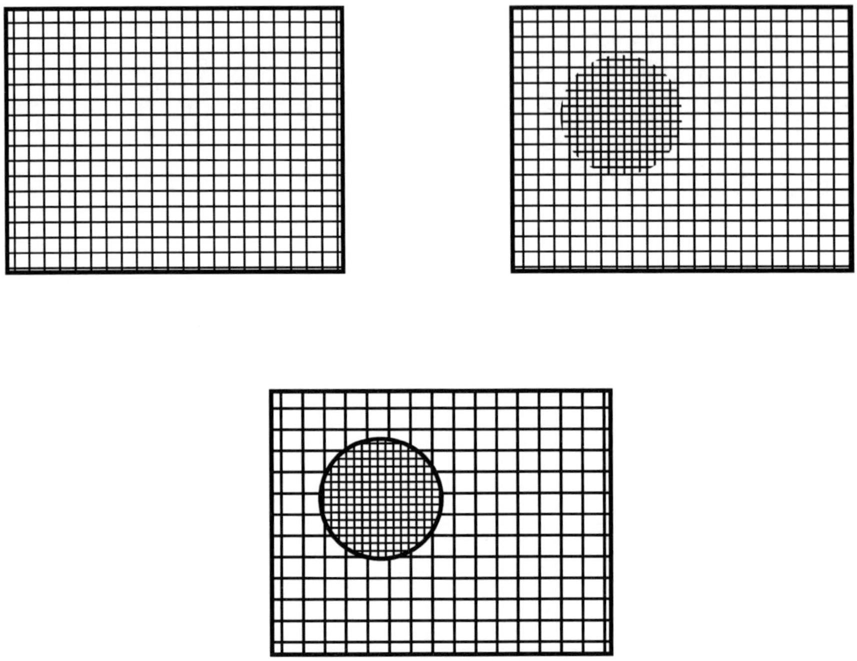

reverberation

Figure 5. Two levels of representation.

Such an approach helps to conciliate formal and semantic approaches to grammar. Some syntactic theories convey formal labels like Nominative case, Subject or Theme: they are language specific notions because neither of these labels is universal. Some other syntactic theories convey semantic labels like Agentive or Experiencer: they are universal notions for they reflect the behaviour of actants in the external word. No adequate theory of language may go without any of them: if it lacks the former, then it will not fulfil the requirement of linguistic empirical adequacy; if it lacks the latter, then it will fail the requirement of referential adequacy. Liminar grammar (LG) is a theory of language that has benefited from many of Thom's epistemological assumptions (López-García 1990). Thus, like catastrophe theory, liminar grammar conceives its object of study as a topological space that is constituted by linguistic signs together with their metalinguistic descriptions. Moreover, liminar grammar establishes a correspondence between topological concepts and gestaltic ones when it states that the foreground/background distinction of Gestalt Psychology corresponds to the topological opposition of closed sets/open sets.

Nevertheless, while dynamic applications of CT to LG usually succeeded, there was a problem with the archetypal semantics, because it is used to describing linguistic scenes that each language formalizes on its own no matter if they all relate to a unique world situation. The ignition/reverberation opposition would allow LG to adopt a catastrophe theoretical frame also in language-specific utterances that are described by means of the figure-ground/boundary frame. Although actantial roles belong to the referent and cannot be taken as language-specific, verbs certainly are. But a verb is simply a boundary that establishes a transition – a neural reverberation – among several actantial arguments that fire – neural ignition – individually. This explanation does not contradict common sense assumptions as CT sometimes did: nouns are perceptually prior to verbs but it is the verbal meaning that represents a transition among them and that properly speaking maps the external world into linguistic networks.

From a genetic point of view this means that the emergence of protolanguage was preceded by many object images being recorded in memory. But protolanguage is not possible without those images being tied up. Although object images were accumulating in the mind of human predecessors during a long period of time, the web that relies them must have appeared rather suddenly. Hence, we suppose that protolanguage origins took place in two phases following a punctuated equilibrium model (López-García 2005): firstly, nouns were slowly stored in the mind: secondly, actantial schemes, which are motor schemes, rapidly – i.e. catastrophically – got integrated in the protolinguistic frame and language was born. The nominal storage was different in each human group, whereas the motor storage, which related to hand movements and gestures, was universal and supported an actantial archetypal semantics: when the later overlapped the former, it was converted in language-specific too, though its mental root remained as universal as before.

References

Brandt, Per Aage (1984). *Dynamiques du sens*, Aarhus University Press.

Brandt, Per Aage (1991). "Pour une sémiotique de la promesse. Quelques réflexions théoriques", in P. Aa. Brandt et A. Prassoloff (éds.), *Qu'est-ce qu'une promesse?*, Aarhus, Aarhus U.P.

Edelman, Gerald (1987). *Neural Darwinism. The Theory of Neuronal Group Selection*, New Cork, Basic Books.

Edelman, Gerald (1988). *Topobiology. An Introduction to Molecular Embriology*, New Cork, Basic books.

Fuster, Jaume (1995). *Memory in the Cerebral Cortex*, Cambridge, The MIT Press.

Hebb, Donald Holding (1949). *The organization of behavior. A neuropsychological theory*, New York, Wiley.

Jackendoff, Ray (1987). "On Beyond Zebra. The relation of linguistic and visual information", *Cognition*, 26, 1987, 89–114.

López-García, Ángel (1990). *Introduction to Topological Linguistics*, University of Minnesota.

López-García, Ángel (1991). "Análisis de la conversación y teoría de catástrofes", in: *Voz y Letra*, II–1, 3–16.

López-García, Ángel (2005). "Le modèle de l'équilibre intermittent et l'origine du langage", in: *Cuadernos de Filología Francesa 16*, Universidad de Extremadura, 25–37.

Petitot-Cocorda, Jean (1985a). *Les catastrophes de la parole*, Paris, Maloine.

Petitot-Cocorda, Jean (1985b). *Morphogenèse du sens I*, Paris, PUF.

Pulvermüller, Friedemann (2002). *The Neuroscience of Language. On Brain Circuits of Words and Serial Order*, Cambridge, Cambridge University Press.

Thom, René (1972). *Stabilité structurelle et morphogenèse*, Paris, Ediscience.

Thom, René (1977). "La théorie des catastrophes, état présent et perspectives", in Zeeman (ed.), 615–621.

Thom, René (1983). *Paraboles et catastrophes. Entretiens sur les Mathématiques, la Science et la Philosophie* réalisés par Giulio Giorello et Simona Morini, Paris, Flammarion.

Thom, René (1988). *Esquisse d'une sémiophysique. Physique aristotélicienne et théorie des catastropes*, Interéditions, Paris.

Thom, René (1990). *Apologie du logos*, Hachette, Paris.

Thom, René (1992). *Pouvoirs de la forme*, in: F. Stjernfelt (col.), Aarhus Universitet, Center for Kulturforskning.

Wildgen, Wolfgang (1981). "Archetypal Dynamics in Word Semantics. An Application of Catastrophe Theory", in H. J. Eikmeyer and H. Rieser (eds.), *Words, Worlds, and Context*, de Gruyter, Berlin: 234–296.

Wildgen, Wolfgang (1982). *Catastrophe Theoretic Semantics*, John Benjamins, Amsterdam.

Wildgen, Wolfgang (1999). *De la grammaire au discours. Une approche morphodynamique*, Peter Lang, Bern.

Wildgen, Wolfgang (2007). *The evolution of human language: Scenarios, principles, and cultural dynamics*, John Benjamins, Amsterdam.

Zeeman, Cristopher (1977).*Catastrophe Theory. Selected Readings*, Addison-Wesley, Reading MA.

Part Two / Deuxième partie
Chapters in French / Chapitres en Français

“Le hiatus entre le logique et le morphologique”. Prédication et Perception

JEAN PETITOT

1. L’héritage de René Thom

Pour pouvoir évaluer correctement l’héritage de la théorie des catastrophes, il faut, je crois, voir les choses à un double niveau. Il y a d’abord évidemment la postérité directe des idées personnelles de René Thom. Et puis, il y a ensuite le développement d’idées analogues à celles qu’il a été l’un des tout premiers à promouvoir de la façon la plus efficace et la plus profonde, à savoir l’application à des domaines nouveaux d’un certain nombre de formalismes – géométrie différentielle, théorie des singularités, théorie des bifurcations – qui, jusque-là, n’étaient essentiellement appliquées qu’en physique, par exemple dans l’étude des ruptures de symétrie, des phénomènes critiques ou des phénomènes de transitions de phase.

Je me souviens que lorsque, à la fin des années 60, j’ai écris à René Thom – que je connaissais déjà comme l’un des plus grands spécialistes mondiaux de la théorie des singularités – pour lui dire à quel point j’étais enthousiasmé par l’idée d’une approche du sens, du langage et des contenus cognitifs, qu’ils soient perceptifs ou conceptuels, en termes de géométrie différentielle, de théorie des singularités et de théorie des bifurcations, pratiquement personne, en dehors de Roman Jakobson, ne comprenait alors ce qui était en jeu. Dans les années 70, quelques pionniers se sont lancés dans le “morphological turn”: Wolfgang Wildgen et Per Aage Brandt en sémio-linguistique, Giulio Giorello, Fernando Gil et Kristof Pomian pour l’Enciclopedia Einaudi. Il y a eu toutes ces années de l’extraordinaire séminaire du samedi matin à l’IHES de Bures-sur-Yvette et tout cela a abouti au début des années 80 au Colloque de Cerisy *Logos et Théorie des Catastrophes*, au livre de Wolfgang, à ma thèse, puis à la thèse de Per Aage, et au débat sur le structuralisme dynamique à l’intérieur de l’Ecole sémiotique greimasienne. Il y a là une douzaine d’années d’une richesse exceptionnelle dont il faudra un jour raconter l’histoire dans le détail.

A l'époque, d'autres groupes d'avant-garde travaillaient sur des idées voisines: le groupe d'Ilya Prigogine à l'Université Libre de Bruxelles, le groupe de Henri Atlan et Jean-Pierre Dupuy au CREA et le groupe de Synergétique de Haken.

La situation a considérablement évolué vers le mileu des années 80 avec le développement explosif des réseaux de neurones. Les modèles dynamiques se sont alors rapidement imposés partout, essentiellement parce que dans ces modèles les dynamiques internes et externes deviennent explicites. Mais la référence à Thom y était pratiquement inexistante.

En ce qui concerne la postérité *directe* des idées personnelles de René Thom, ses disciples ont beaucoup travaillé. Même si René n'a pas constitué à proprement parler d'école, il a fécondé la pensée d'un nombre considérable de personnes, aussi bien dans le domaine des mathématiques pures que dans celui des applications et celui de la philosophie. Tous les chercheurs qu'il a inspirés ont continué à développer ses œuvres. Mais il est vrai que, sur le plan sociologique et médiatique, après les grands débats des années 70 et 80, la théorie des catastrophes conçue au sens strict a connu un certain retrait, ce qui a parfois pu faire dire à René Thom vers la fin de sa vie qu'elle était un peu un échec sociologique. Mais même à supposer que ce soit effectivement le cas, cet échec relatif se trouve, selon moi, très largement compensé par l'extraordinaire succès des idées du type de celles qu'il a développées.

En particulier dans le domaine des sciences cognitives et sémio-linguistiques, les idées introduites par Thom et Zeeman à la fin des années 60, à savoir qu'il est pertinent d'utiliser la théorie des attracteurs de systèmes dynamiques non linéaires et la théorie des bifurcations de ces attracteurs pour comprendre l'activité mentale et la structure des représentations mentales, ces idées qui étaient extraordinairement nouvelles et provocatrices à l'époque, sont devenues maintenant des idées-forces très largement admises. On peut même dire que, dans des domaines comme l'étude de la perception et des relations entre le langage et la perception, le paradigme dynamique est devenu dominant. Or il est profondément thomien même si tous les spécialistes qui y travaillent ne font pas systématiquement référence à René Thom. C'est pourquoi je pense que l'on peut dire que dans les domaines des sciences humaines qui intéressaient René Thom, sa théorie morphodynamique des formes et des structures s'est considérablement développée et a été confirmée par de nombreuses données expérimentales. En ce sens, il y a en définitive une très grande réussite des idées thomiennes.

Et cela est vrai y compris dans le domaine de la morphogenèse biologique où René Thom estimait avoir introduit ses idées les plus fondamentales et où les controverses ont été les plus chaudes. Là aussi, comme pour les sciences humaines, on peut dire que l'utilisation des modèles dynamiques, de la théorie des bifurcations engendrant des discontinuités qualitatives dans des substrats physiques, s'est beaucoup développée. On pourrait y ajouter d'autres modèles, par exemple les modèles de morphogenèse de type Turing fondés sur des équations différentielles de réaction-diffusion.

Ce qui a peut-être le plus changé d'avec la façon dont René Thom lui-même voyait les choses, c'est qu'il semble désormais possible d'établir une convergence avec les résultats de la génomique. L'on commence à comprendre le contrôle génétique du développement, et l'analyse de ces gènes permet de découvrir comment les réactions morphogénétiques qui intéressaient René Thom peuvent être contrôlées par le génome. René Thom a eu, comme on le sait, des polémiques assez vives avec la biologie moléculaire, mais je pense que le conflit est en train d'être dépassé.

2. La prédication selon René Thom

Parmi tous les sujets de sémiolinguistique qu'il serait possible d'aborder dans ce colloque d'hommage, je vais privilégier celui de la *prédication* car on en parle peu alors qu'il est pourtant fondamental. En effet, en ce qui concerne les *jugements perceptifs*, il existe ce que Thom appelait un "hiatus infranchissable entre le logique et le morphologique" (Thom 1988, p. 248). Cela est dû au fait que dans l'expression "jugement perceptif" *les deux termes renvoient à des univers formels complètement différents:* "jugement" à une syntaxe logique de catégories grammaticales, "perceptif" à une géométrie morphologique.

Or les approches cognitivistes du langage ont précisément conduit (entre autres) à repenser en profondeur les liens entre les structures syntactico-sémantiques des langues naturelles et les structures de la perception. Elles ont fait apparaître un nombre considérable de nouveaux problèmes dont l'approche formelle exige des outils appropriés assez éloignés de ceux, classiques, de la sémantique formelle.

J'aimerais, dans cette perspective, proposer quelques remarques concernant les formes les plus primitives de prédication dans les jugements perceptifs comme "la neige est blanche" (Tarski), "l'hexaèdre est rouge" (Husserl), "le ciel est bleu" (Thom). Le caractère tautologique de la définition tarskienne de la vérité: l'énoncé "la neige est blanche" est vrai si et seulement si l'état de choses "la neige est blanche" est réalisé, est le symptôme de la mise entre parenthèses de la perception dans l'analytique logique classique. C'est sur ce point que je voudrais axer ma réflexion.

Il y a déjà fort longtemps que je m'intéresse à ce sujet. En particulier pour le 16th International Wittgenstein Symposium de 1994, *Philosophy and Cognitive Science*, j'avais déjà introduit un certain nombre de propositions pour *formaliser les jugements perceptifs* au moyen de la théorie des topoï[1]. Ce sont ces idées que j'aimerais préciser ici.

2.1. Prédication et valeur de vérité des jugements perceptifs

Pour obtenir une formalisation morphologique, reprenons la façon dont, dans l'*Esquisse de Sémiophysique* (Chapitre 6. La dynamique aristotélicienne comme sémiophysique, F. Axiomes de l'acte, 2. Phrases univalentes), Thom analyse un jugement perceptif de base comme "le ciel est bleu". Le remplissement de l'extension spatiale W d'un objet S par une qualité sensible (comme une couleur) p appartenant à un genre de qualité G se décrit par une application:

$$\begin{array}{rcl} g : W & \rightarrow & G \\ w & \rightarrow & g(w) \end{array}$$

qui, à tout point w de W, associe la valeur de la qualité en ce point. En fait, il est naturel d'interpréter g comme *une section de la fibration* $\pi : E = W{\times}G \rightarrow W$ de base W et de fibre G qui associe au couple (w, s) d'un point w de la base W et d'un point s de la fibre G le point w. Une section d'une fibration $\pi : E \rightarrow W$ est une application $\sigma : W \rightarrow E$ qui "relève" π, i.e. telle que $\pi \circ \sigma = 1_W$. La valeur $g(w)$ de g en w est donc considérée comme appartenant à l'exemplaire de G qui est la fibre $\pi^{-1}(w)$ de π au-dessus de w.

La nécessité d'introduire une fibration est fortement soulignée par Thom:

> C'est pourquoi, en tant que mathématicien, pour définir l'homéomère, je dois multiplier l'espace substrat par un espace invisible, un espace (interne) de qualités – un espace de genre –, pour y définir le bord de ma qualité homéomère.[2]

1 Petitot (1994a).

2 Dans la biologie d'Aristote, les homéomères sont les parties homogènes des substrats.

Dans une certaine mesure, toute la théorie des catastrophes a développé cette idée de base que la façon dont les qualités "secondes" remplissent les extensions spatio-temporelles ("qualité première") devait s'interpréter, comme en physique, par des sections de fibrations bien choisies.

Le *remplissement* de S par la qualité p se trouve ainsi décrit par la section $g : W \rightarrow W \times G$ de π qui associe à tout point w de W, le couple $(w, g(w))$ où $g(w)$ est la valeur de la qualité p en w. La section g décrit donc le remplissement, que nous noterons $<S, p>$, autrement dit la structure perceptive (évidemment hypersimplifiée).

Il est important de noter que le concept géométrique de fibration est neuro-physiologiquement pertinent. Les structures (hyper)columnaires du cortex (qui permettent d'associer à chaque position rétinienne un élément comme une couleur, une texture, une direction, etc. et donc de définir des *champs rétinotopiques* de tels éléments) constituent des implémentations (évidemment discrétisées) de fibrations.

Ce formalisme morphologique permet de schématiser les relations de dépendance unilatérale (les hiérarchies) entre les genres. Le fait que l'extension spatiale (qualité "première") soit ontologiquement première par rapport aux qualités "secondes" se trouve traduit par le fait que, dans la fibration $\pi : E \rightarrow W$, l'extension W est *la base* (qui peut exister de façon indépendante) alors que le genre de qualités G est *la fibre*.

En général l'espace de genre G est en plus *catégorisé* en espèces (en "essences" qualitatives ou *species*, par exemple les *types* $C_1, ..., C_n$ de couleurs) par un potentiel $V: G \rightarrow \mathfrak{R}$.[3] Les bassins d'attraction des minima définissent alors les *catégories* (les espèces) de G et les minima fonctionnent comme des *prototypes*. Si $p \subset G$ est une de ces catégories, si ∂p est son bord dans G et si S est le nom de l'entité de substrat W considérée, le jugement perceptif "S est p" s'interprète selon Thom par le fait **que l'image $g(W)$ de la section $g : W \rightarrow W \times G$ est encapsulée dans le cylindre $W \times \partial p$** (cf. Thom 1988, pp. 158 sq et 205 sq) : "S est p" est vrai ssi $g(W) \subset W \times \partial p$.

> Il s'agit de la capture de l'image de la section $g : W \rightarrow W \times G$ dans l'espace d'un genre G par une dynamique interne à G (conduisant à un minimum attracteur).
>
> On considérera qu'à tout genre – entité seconde – est associé un substrat G, et sur cet espace on aura un espace de qualité (état interne). Alors G est défini comme la réunion des bassins d'un potentiel $V : G \rightarrow \mathfrak{R}$; le bassin d'un minimum est le substrat d'une espèce (*eidos*). Par exemple, sur l'espace continu des impressions de couleur (espace à trois di-

3 $\mathfrak{R}$ est le corps des nombres réels.

mensions $\Re^3$ selon la théorie classique), l'adjectif bleu serait défini par un bassin b, de bord ∂b. La signification d'une phrase attributive comme "Le ciel est bleu" peut être ainsi géométrisée. Le substrat du "ciel" est une demi-sphère céleste D^2; du point de vue de la couleur, l'état du ciel est défini par une section $\sigma : D^2 \rightarrow \Re^3 \times D^2$ du morphisme $\Re^3 \times D^2 \rightarrow D^2$. Le "ciel est bleu" est l'acte signifiant que cette section est "capturée" par le bassin b, autrement dit la section $\sigma : D^2 \rightarrow \Re^3 \times D^2$ est tout entière contenue dans le tube $b \times D$ avec le bord duquel ($\partial b \times D$) elle est "entrelacée".

Nous avons donc deux CNS (conditions nécessaires et suffisantes) de validité pour un énoncé "*S* est *p*":

1. La CNS tarskienne: l'énoncé "*S* est *p*" est vrai ssi l'état de choses correspondant "*S* est *p*" est vérifié ("*S* est *p*" ne pouvant d'ailleurs être représenté linguistiquement qu'au moyen du jugement "*S* est *p*").
2. La CNS thomienne: l'énoncé "*S* est *p*" est vrai ssi l'image $g(W)$ de la section $g : W \rightarrow W \times G$ est encapsulée dans le cylindre $W \times \partial p$.

La première est logique et prédicative mais "aveugle" (sans intuition remplissante). La seconde est morphologique mais ante-prédicative et préjudicative. Comment en effectuer la synthèse? Comment penser le lien entre le schématisme géométrique de la perception et la formalisation traditionnelle des jugements en termes de logique formelle? Le problème est délicat car, comme l'avait déjà profondément remarqué Wittgenstein, dans des jugements perceptifs du type "*S* est *p*", "*X* est plus clair que *Y*", etc., *la spatialisation des qualités n'est pas explicite.*

Dans la CNS de Thom, la vérité est *spatialement localisée* et c'est tout le problème du "hiatus entre le logique et le morphologique". C'est cela qu'il faut comprendre. Pour avancer faisons d'abord un détour par Husserl.

3. Présentation perceptive et représentation propositionnelle dans *Erfahrung und Urteil* de Husserl

3.1. Présentation géométrique et représentation propositionnelle

Les formules atomiques "*S* est *p*" que sont les jugements d'attribution de qualités sensibles à des objets spatialement localisés se situent au degré zéro de l'échelle logique. Tant leur syntaxe que leur sémantique sont triviales.

Mais cette trivialité évidente disparaît dès que l'on essaye de comprendre les liens qui peuvent exister entre leur structure syntactico-sémantique et la scène perceptive – que nous noterons <*S*, *p*> – qu'ils décrivent et dénotent.

La scène perceptive <*S*, *p*> correspond à la donnée d'un domaine spatial W_S (l'extension de l'objet *S*) rempli par une qualité sensible *p*. Elle est "synthétique", de format géométrique et "présentationnelle" (au sens de l'opposition philosophique et phénoménologique classique entre *Darstellung* (présentation) et *Vorstellung* (représentation)). Au contraire, le jugement prédicatif "*S* est *p*" est quant à lui "analytique", de format propositionnel et "représentationnel". Quel rapport il y a-t-il entre ces deux formatages, c'est-à-dire, entre d'un côté le remplissement intuitif d'extensions spatiales par des qualités et d'un autre côté les catégories syntaxiques de la prédication?

Ces problèmes ont été fort peu étudiés par les traditions logico-sémantiques. Il existe pourtant une exception notable, de première importance, celle de Husserl. En particulier dans les extraordinaires réflexions sur l'origine perceptive de la prédication que l'on trouve dans *Erfahrung und Urteil*.

3.2. *Erfahrung und Urteil*

Dans cet ouvrage posthume[4] (1939, Husserl ayant disparu en 1938), édité à Prague par son disciple et assistant Ludwig Landgrebe et dont le sous-titre *Untersuchungen zur Genealogie der Logik* exprime bien les intentions, Husserl cherche à clarifier phénoménologiquement les origines de la prédication. Il y élabore (entre autres) une théorie des propositions atomiques "*S* est *p*" dans le cadre d'une apophantique formelle (i.e. d'une théorie syntaxique) et d'une ontologie formelle (i.e. d'une théorie sémantique comme la méréologie ou la théorie des ensembles). Si sa perspective est "généalogique", c'est parce que, selon lui, la logique formelle classique occulte le problème fondamental de *l'évidence* dans le concept logique de vérité. "Evidence" signifie ici l'immédiateté de la donation des phénomènes dans la présentation perceptive (l'"immediate acquaintance" des russelliens).

> Le caractère formel de l'analytique logique consiste en ce qu'elle ne s'interroge pas sur la qualité matérielle [i.e. perceptive] de ce quelque chose [donné dans la perception], qu'elle n'envisage les substrats qu'en fonction de la forme catégoriale qu'ils prennent dans le jugement. (p. 28)

---

4 Husserl (1954).

Comme "jugement catégorique fondé dans la perception" (p. 79), un jugement prédicatif à contenu perceptif de type "*S* est *p*" est enraciné dans l'expérience ante-prédicative et pré-judicative du monde perceptivement donné et c'est bien cet enracinement – ce qu'il appelle une "fondation" – que Husserl veut thématiser comme tel. Il retrace pour ce faire la "genèse catégorielle" des catégories logiques "primitives" (sujet/prédicat), genèse qui convertit l'unité perceptive synthétique <*S, p*> de l'extension spatiale d'un substrat *S* délimitée par un bord (contour apparent) et possédant un moment dépendant *p* (qualité sensible) en l'unité syntaxique analytique de la proposition "*S* est *p*". Il ramène cette conversion d'une présentation perceptive en une représentation propositionnelle à une opération réflexive de *thématisation* et de *typification logique* qui typifie les substrats en sujets et les moments dépendants en prédicats. Telle est selon lui: "l'origine des premières catégories dites 'catégories logiques'". (p. 134)

Comme il y insiste,

> dans le jugement prédicatif le plus simple, une *double information* est traitée (p. 252)

car *sous* l'information syntaxique catégoriale "sujet/prédicat" concernant les "formes fonctionnelles" des termes de la proposition, il existe une autre information concernant les "formes noyaux" /substrat = indépendance/ et /moment qualitatif = dépendance/. Selon Husserl, la prédication est un processus basé sur

> le recouvrement des formes noyaux comme matériel syntaxique pour les formes fonctionnelles. (p. 252).

C'est cette typification logique en catégories syntaxiques des relations synthétiques de dépendance substrats-qualités qu'il s'agit de formaliser.

3.3. Pertinence d'une approche morphologique thomienne

Si l'on veut, au-delà de la description eidétique pure, transformer la description phénoménologique en source de modélisation, il faut alors pouvoir formaliser les phénomènes de remplissement qualitatif constitutifs des états de choses <*S, p*>. *Or c'est précisément cela que permet l'approche morphologique de René Thom.*

Cette possibilité de disposer d'une *présentation morphologique d'un schème sensible* <*S, p*> qui soit différente de la proposition "*S* est *p*" est absolument nécessaire si l'on veut briser

(i) le cercle vicieux qui affirme que les structures et les propriétés qualitatives du monde sensible n'existent qu'à travers les *jugements* qui les interprètent et ne peuvent pas être montrés (présentés) autrement (c'est-à-dire au fond que toute *Darstellung* est toujours-déjà une *Vorstellung*), et

(ii) le caractère tautologique de la définition tarskienne de la vérité: l'énoncé "la neige est blanche" est vrai ssi l'état de choses "la neige est blanche" est réalisé.[5]

Elle permet aussi d'entrevoir la solution à un certain nombre de difficultés qui ont fait jusqu'ici l'objet d'un nombre considérable de discussions philosophiques.

(i) La représentation propositionnelle "*S* est *p*" de la présentation morphologique <*S, p*> possède un contenu qui est *intensionnel* à un double titre. D'abord parce qu'il représente l'état de choses sous un certain aspect, celui présenté par <*S, p*>. Ensuite parce que les spécificités de l'aspect de <*S, p*> (par exemple l'extension exacte de *S*, les valeurs exactes de *p*) *ne sont pas* réflétées dans le jugement "*S* est *p*": celui-ci fonctionne en fait de façon *indexicale*. Tout jugement perceptif élémentaire ne peut fixer sa référence que de façon pragmatique: le lien entre l'énoncé "*S* est *p*" et la scène perceptive <*S, p*> est contrefactuel. Comme nous allons le voir, la sémantique naturelle des jugements perceptifs est donc une sémantique intensionnelle *à la Kripke*, mais d'un nouveau type.

(ii) Cela explique le fait que la représentation "*S* est *p*" possède une propriété sémantique en vertu de laquelle elle représente le schème sensible <*S, p*>. La relation entre *S* et *p* dans le jugement "*S* est *p*" est donnée par la relation de dépendance constitutive de la structure morphologique du schème sensible <*S, p*>. Or ce dernier diffère de l'état de choses objectif associé (un schème sensible n'est pas objectif au sens "chosique"). D'où la possibilité de méprise représentationnelle.

---

5 Pour une critique de cette définition tautologique dans le cadre de la théorie des topoï que nous utilisons plus bas, cf. Moerdijk-Reyes (1991).

4. L'indexicalité pragmatique des jugements perceptifs

Les jugements perceptifs sont tous d'une certaine façon *indexicaux*. La spatialité implicite des objets (la localisation et l'extension) y fonctionne comme fonctionnent les déictiques: elle doit être actualisée par la situation perceptive qui joue en quelque sorte le rôle d'un contexte pragmatique. On peut dire aussi *que l'espace modalise la vérité des jugements perceptifs*. Dans un jugement d'attribution de qualité de type "S est p" on ne dit pas que S est un symbole de variable référant à un individu possédant la propriété p. On dit que S réfère à un individu d'extension W et que W est "rempli" par la qualité p. Les conditions de vérité du jugement présupposent donc que S réfère à une section $g(W)$ de la fibration associée à la qualité p. Mais l'extension W demeurant implicite dans la syntaxe de l'énoncé et n'intervenant qu'au niveau de sa sémantique, on peut dire qu'elle fonctionne de façon indexicale.

5. La pertinence des concepts de Faisceau et de Topos

L'on voit que si l'on articule ainsi morphologie perceptive et logique judicative *en respectant leur altérité*, alors le problème devient de formaliser l'analyse husserlienne-thomienne des liens entre un schème morphologique $<S, p>$ (intuition remplissante) et un jugement "S est p" (intention de signification) et, donc, de faire droit à la priorité du formatage morphologique (présentationnel) sur le formatage propositionnel (représentationnel) pour pouvoir ensuite comprendre la conversion faisant passer de l'un à l'autre.

Essayons d'imaginer un chemin menant du signal optique à un jugement en essayant de répérer exactement à quel moment géométrie et logique doivent s'articuler.

1. Le signal doit d'abord être traité, régularisé et segmenté. On connaît des algorithmes très performants: ondelettes de Mallat, modèle variationnel de Mumford-Shah, équations de diffusion anisotropes de Morel, etc.[6]

6 Cf. par exemple Petitot (1994b) et (2003b).

2. Les aspects, les profils (Abschattungen, adumbrations) – et en particulier les contours apparents – doivent être reconnus comme des aspects d'objets. On sait que René Thom s'était énormément intéressé à ce problème. C'est une très jolie application de la théorie des singularités. En se situant dans la grassmannienne des projections on associe un contour apparent (i.e. un lieu singulier avec des singularités) à chaque point de vue, et la question de savoir comment les types génériques de contours apparents stratifient l'espace des points de vue est un problème difficile encore non résolu.

3. Mais ce qui doit être transformé en jugement est d'abord un aspect particulier, celui sélectionné par l'obervateur. Comme nous l'avons vu, la sélection du point de vue et son changement correspondent à la nature *pragmatique* des jugements perceptifs.

4. C'est donc le remplissement d'une extension spatiale par une qualité sensible qui constitue bien la donnée primitive, ainsi que René Thom l'avait tout de suite formalisée avec la notion de point régulier et de point singulier dans *Stabilité structurelle et Morphogenèse*. C'est bien elle qu'il faut faire passer du géométrique au logique.

L'idée est alors d'utiliser certains résultats fondamentaux concernant les liens entre géométrie et logique qui ont été découverts dans le cadre de la théorie des catégories (au sens mathématique du terme) et, plus précisément, dans celui de la dite *théorie des topoï*. Très intuitivement, les idées de bases sont les suivantes.[7]

(i) Les remplissements possibles de domaines spatiaux W d'un espace ambiant M par des qualités sensibles – et donc les sections σ qui les modélisent – reposent sur une dialectique *du local et du global* (possibilité de restreindre un recouvrement à un sous-domaine, possibilité de recoller des recouvrements compatibles, etc.) qui est caractéristique de ce que l'on appelle la structure de *faisceau* sur un espace de base M.

(ii) Les opérations formelles catégoriques que l'on peut faire sur des faisceaux – qui sont des objets *géométriques* – sont *exactement parallèles* aux opérations *syntaxiques* que l'on peut faire sur des *symboles* – qui sont des objets *logiques*. C'est cela la grande découverte: elle est due à

7 Pour une introduction à la théorie des topoï, cf. Asperti-Longo (1991) et MacLane-Moerdijk (1992).

William Lawvere à la fin des années 60. On dit que la catégorie des faisceaux sur un espace de base *M* possède la structure de *topos*.

(iii) On peut par conséquent associer à la catégorie des faisceaux sur *M un langage formel* (ce que l'on appelle sa "logique interne"). Dans ce langage logique une variable *x* va être interprétée *syntaxiquement* par un faisceau *X* (par exemple le faisceau des sections d'une fibration $\pi : M \times G \to M$) qui représente *son type logique*. Mais, *sémantiquement*, *x* sera interprétée comme une *section* particulière de *X* définie sur un domaine particulier *W* de *M.* [8] Cette sémantique très particulière s'appelle la sémantique de Kripke-Joyal des topoï.

Autrement dit, le formalisme "toposique" permet de typifier logiquement des symboles qui réfèrent à des remplissements de domaines spatiaux par des qualités.

C'est exactement le genre de formalisme dont on a besoin pour formaliser la description eidétique de *Erfahrung und Urteil* à partir de la géométrisation de la prédication proposée par Thom.

Il s'agit donc de reprendre l'analyse logique des propositions (et donc les bases mêmes de la sémantique) à partir de ce primat du perceptif. Il s'agit de relier au moyen d'une logique géométrique une typification logico-catégorielle et un schématisme morphologique. Comme le formalisme logique, le schématisme morphologique est une structure idéale se réalisant dans des actes et des processus mentaux. Mais son idéalité est "esthésique" (au sens de l'Esthétique transcendantale kantienne) et non pas symbolique.

6. Faisceaux, Topoï et Logique

6.1. Remarques préliminaires

1. La théorie des topoï a été inventée et développée par Alexandre Grothendieck et son école dans les années 60 pour résoudre des problèmes très sophistiqués de géométrie algébrique (construction de théories co-

8 Techniquement, il faut tenir compte du fait que, traditionnellement, les sections de faisceaux sont définies sur des ouverts de l'espace topologique de base. Mais on peut généraliser cette situation traditionnelle (cf. section VI).

homologiques généralisées). Il peut donc paraître étrange de vouloir l'utiliser pour des problèmes cognitifs non mathématiques et, qui plus est, apparemment élémentaires. Mais en fait si ces problèmes sont traités de façon neurocognitive (ce qui m'intéresse au premier chef mais dont je ne parle pas ici), ils ne sont plus du tout triviaux et doivent donc être modélisés avec des outils convenables.

2. La théorie des catégories est certainement la meilleure ontologie formelle dont nous disposions actuellement (par beaucoup d'aspects elle est supérieure à la théorie des ensembles) et il est donc naturel de s'y placer.

3. Certains spécialistes de la théorie des topoï ont déjà appliqué ces outils à la clarification de certains problèmes sémantiques. Par exemple Gonzalo Reyes (qui a travaillé, avec Eduardo Dubuc, Anders Koch, Ieke Moerdijk et Marta Bunge sur les applications de la théorie des topoï à la *Géométrie différentielle synthétique*) a utilisé ces techniques pour formaliser la théorie kripkéenne des noms propres comme désignateurs rigides.

4. En général, la théorie des topoï est utilisée en logique formelle et en informatique théorique comme un outil pour le λ-calcul typé (cf. par exemple les travaux de John Mitchell, Philip Scott, Giuseppe Longo, etc.). Comme nous allons le voir, quand des variables sont typées par des objets d'un topos, les propriétés catégoriques du topos conduisent à une logique interne (en général intuitionniste) qui est un λ-calcul typé. Dans la mesure où la correspondance de Curry-Howard montre que les formules peuvent être traitées comme des types, les preuves comme des λ-termes et la réduction d'une preuve par élimination de coupures à la réduction d'un λ-terme à sa forme normale, la théorie des topoï est devenue un outil fondamental pour comprendre la sémantique des langages formels, et en particulier des langages de programmation.

Dans ces applications logiques, l'origine (la "généalogie") géométrique des concepts de faisceau et de topos n'est pas pertinente. Ici, nous utilisons au contraire cette origine géométrique de façon essentielle pour clarifier la montée cognitive du niveau morphologique-perceptif vers le niveau propositionnel-prédicatif.

6.2. Le concept de faisceau

Pour le concept de faisceau, le concept topologique primitif n'est plus celui de point mais celui *d'ouvert* (ce que Peter Johnstone appelle la "pointless topology").[9]

De façon abstraite, une fibration sur un espace de base M est caractérisée par l'ensemble de ses sections $\Gamma(U)$ sur les ouverts $U \subset M$. C'est un dispositif produisant des sections. Si $s \in \Gamma(U)$ est une section sur U et si $V \subset U$, on peut naturellement considérer la *restriction* $s_{|V}$ de s à V. Cette restriction est une application $\Gamma(U) \rightarrow \Gamma(V)$. Il est évident que si $V = U$ alors $s_{|V} = s$ et que si $W \subset V \subset U$ et $s \in \Gamma(U)$ alors $(s_{|V})_{|W} = s_{|W}$ (transitivité de la restriction). On obtient ainsi un *foncteur contravariant* $\Gamma : \mathcal{O}^*(M) \rightarrow$ **Ens** de la catégorie $\mathcal{O}(M)$ des ouverts de M dans la catégorie **Ens** des ensembles (les objets de $\mathcal{O}(M)$ sont les ouverts de M et les morphismes sont les inclusions d'ouverts).

Réciproquement, soit Γ un tel foncteur – ce que l'on appelle un *préfaisceau* sur M. Pour avoir une chance d'être le foncteur des sections d'une fibration, il est clair que Γ doit satisfaire les deux propriétés caractéristiques suivantes:

(F_1) Deux sections localement égales sont globalement égales. Si $\mathcal{U} = (U_i)_{i \in I}$ est un recouvrement ouvert de la base M et si $s, s' \in \Gamma(M)$ sont deux sections globales, alors, si $s_{|U_i} = s'_{|U_i}\ \forall i \in I$, on a l'égalité $s = s'$.

(F_2) Une famille de sections s_i qui sont définies sur un recouvrement ouvert $\mathcal{U} = (U_i)_{i \in I}$ d'un ouvert U et compatibles sur les intersections $U_i \cap U_j$ peut être recollée en une section globale s sur U. Autrement dit si $s_i \in \Gamma(U_i)$ est une famille sur $\mathcal{U} = (U_i)_{i \in I}$ et si les s_i sont compatibles i.e. si $s_{i|U_i \cap U_j} = s_{j|U_i \cap U_j}$ quand $U_i \cap U_j \neq \varnothing$, alors $\exists s \in \Gamma(M)$ telle que $s_{|U_i} = s_i\ \forall i \in I$.

Il est remarquable que (F_1) et (F_2) puissent être exprimées de façon purement catégorique. Cela montre en effet que certains des caractères les plus "synthétiques" de l'espace peuvent être décrits dans le cadre de l'ontologie formelle qu'est la théorie des catégories. Par exemple (F_2) dit que la flêche e qui associe à la section s la famille de ses restrictions

9 Dans les sections techniques qui suivent nous traiterons les extensions d'objets comme des ouverts car les concepts géométriques qui nous intéressent ont été définis et étudiés à partir de ce concept primitif. Mais, comme nous l'avons dit, une telle hypothèse n'est pas réaliste puisque la plupart des objets étant délimités par des bords leur extension est fermée.

$$e : s \rightarrow \{s|_{U_i}\}_{i \in I}$$

est l'*equalizer*

$$\Gamma(U) \overset{e}{\rightarrow} \prod_i \Gamma(U_i) \underset{q}{\overset{p}{\rightrightarrows}} \prod_{i,j} \Gamma(U_i \cap U_j)$$

des deux projections p, q correspondant aux inclusions $U_i \cap U_j \subset U_i$ et $U_i \cap U_j \subset U_j$.

Prises comme axiomes pour des préfaisceaux, les propriétés (F_1) et (F_2) définissent une structure plus générale que celle de fibration, nommément celle de *faisceau.* On peut d'ailleurs montrer que si les axiomes (F_1) et (F_2) sont satisfaits alors on peut *représenter* le foncteur section Γ par une structure fibrée généralisée $\pi : E \rightarrow M$ (appelée un espace "étalé") de telle façon que $\Gamma(U)$ devienne l'ensemble des sections de π sur U. Mais π n'est plus nécessairement localement triviale comme doit l'être par définition une fibration. La fibre E_x de E en x est la limite inductive:

$$= \lim_{V \subset U \in \mathcal{U}_x} \{(\Gamma(U), \Gamma(V \subset U)\}$$

(où $\mathcal{U}_x$ est le filtre des voisinages ouverts de x). E_x est l'ensemble des *germes* s_x des sections en x. E est la somme des E_x. Si $s \in \Gamma(U)$, elle peut être interprétée comme l'application $x \in U \rightarrow s_x \in E_x$. La topologie de E est alors définie come la plus fine rendant toutes ces sections continues.

Il faut se convaincre que le concept de faisceau n'est pas seulement technique et formalise des propriétés d'essence – eidétiques, "synthétiques a priori" – de l'intuition pure spatiale. Il est *a priori* adapté à toutes les situations où

(i) le formatage spatial se ramène essentiellement à un passage de domaines locaux à des domaines globaux par *recollement*, et

(ii) les entités considérées proviennent d'opérations de *remplissement* (filling-in, *Erfüllung*) de tels domaines spatiaux.

C'est le cas dans de très nombreux domaines de la géométrie et de la physique mathématique et c'est pourquoi ce concept y est devenu omniprésent. *Mais tel est aussi le cas de la perception.* Le recollement y intervient de façon constitutive, et cela à un double titre. D'abord parce que le champ visuel lui-même est constitué par le recollement des champs récepteurs des cellules ganglionnaires[10]. Ensuite parce que l'espace global s'obtient par

10 Cf. par exemple Petitot (2003a).

recollements de différents exemplaires du champ visuel contrôlés kinesthésiquement (par les mouvements des yeux, de la tête et du corps). Contrairement à ce qui se passe en géométrie différentielle classique, la notion de recollement n'est pas imposée ici par la nécessité de prendre en compte des fibrations non globalement triviales. Elle est imposée par le cablage de l'architecture (le hardware neuronal).

Quant au remplissement, il correspond dans les aires rétinotopiques du cortex visuel à des fibrations: le système s'obtient en "recollant" par des connexions latérales cortico-corticales les "hypercolonnes" qui, au-dessus de chaque position rétinienne implémentent la fibre appropriée (direction, couleur, texture, etc.)[11].

6.3. Généralisations

Il existe de nombreuses généralisations de cette situation de base. La plus connue est celle des topologies de Grothendieck: les recouvrements ouverts y sont définis en termes de "cribles" et le concept de faisceau est généralisé en celui de "site".

Une autre généralisation est celle des "*frames*" et des "*locales*". Les "frames" sont des treillis possédant les propriétés des treillis d'ouverts $\mathcal{O}(X)$, autrement dit des treillis distributifs complets possédant des infs (intersections) finis et des sups (unions) quelconques. Les "locales" sont les objets de la catégorie duale. Dans le cas des espaces topologiques la dualité consiste à associer à une application continue $f : X \to Y$ l'application $f^* : \mathcal{O}(Y) \to \mathcal{O}(X)$, $V \in \mathcal{O}(Y) \to f^*(V) = f^{-1}(V) \in \mathcal{O}(X)$, l'image réciproque $f^{-1}(V)$ d'un ouvert V de Y étant un ouvert de X.

Les points sont alors définis comme des morphismes $x : A \to 2 = \mathcal{O}(1)$ (dans le cas des espaces topologiques, si $A = \mathcal{O}(X)$, les points correspondent à des vrais points $x : 1 \to X$). Soit $Pt(A)$ l'ensemble des points x de A. Une topologie est définie sur $Pt(A)$ par les sous-ensembles:
$\varphi(a) = \{x \in Pt(A) : x(a) = 1\}$. $\mathcal{O}(Pt(A))$ est la meilleure approximation possible du locale A par un locale "spatial".

11 Cf. Ibid.

6.4. Pertinence du concept de topos

Nous nous sommes restreint au cas le plus élémentaire, celui de relations de recouvrement de domaines spatiaux W par des qualités appartenant à un espace de qualités G. Nous avons vu que ces relations sont géométriquement décrites par des sections de fibrations appropriées.

Les faisceaux de sections prennent en charge l'aspect morphologique du schématisme thomien de la prédication. Mais qu'en est-il de son aspect logique? Nous avons vu que le problème est le suivant. Les jugements portant sur des situations de recouvrement d'un domaine W par des qualités appartenant à un genre G ne font pas intervenir dans leur forme syntaxique la localisation de ces qualités. Ils se situent au niveau de G et non pas au niveau de W. La localisation reste *implicite*, c'est-à-dire potentielle. Pour rendre compte de ce caractère potentiel, il faut donc considérer *tout* le faisceau des sections Γ des sections de la fibration $\pi : W \times G \rightarrow W$ *et le traiter en tant que tel comme une unité syntaxique*. Mais dans le même temps, pour qu'un tel jugement puisse être vrai ou faux, il faut – si l'on veut pouvoir appliquer la CNS de Thom – qu'au niveau *de la sémantique* l'extension des objets considérés deviennent *explicite*, autrement dit s'actualise.

La réponse naturelle à ces difficultés est de considérer les faisceaux comme des *types* de variables dont les référents sont des *sections* particulières. Or il se trouve que cela est rendu techniquement possible par un lien profond entre la théorie des faisceaux et la logique qui a été découvert par William Lawvere au début des années 70.

Très brièvement exposées, les idées directrices sont les suivantes. La catégorie $\mathfrak{F}$ des faisceaux sur un espace topologique M possède un certain nombre de propriétés catégoriques fondamentales qui lui confère la structure de *topos*. Or celle-ci *est précisément la structure catégorique permettant d'interpréter un langage des prédicats dans une catégorie d'objets*. Selon Lawvere, cela permet de considérer $\mathfrak{F}$ comme un univers du discours dont les objets *sont des entités variables dépendant d'une localisation spatiale U*. Cette dépendance spatiale

(i) est constitutive des valeurs de vérité et possède donc une pertinence sémantique, mais

(ii) elle n'est pas directement visible dans la syntaxe (qui ne concerne que les faisceaux comme types logiques) et n'est donc pas syntaxiquement pertinente.

Or c'est exactement d'un tel formalisme dont nous avons besoin pour formaliser la façon "indexicale" et "pragmatique" dont opère l'extension spatiale dans les jugements perceptifs.

6.5. Eléments de théorie des topoï

Par définition, un topos $\mathfrak{F}$ est d'abord une catégorie *cartésienne fermée*: il possède des produits fibrés (des pull-backs), un objet terminal – classiquement noté 1 – et des *objets exponentiels* B^A qui permettent *d'internaliser* les ensembles de morphismes $\mathrm{Hom}_{\mathfrak{F}}(A, B)$. Le faisceau B^A est défini de façon évidente en utilisant les restrictions $A|_U$ et $B|_U$ aux ouverts: $B^A(U) = \mathrm{Hom}_{\mathfrak{F}}(A|_U, B|_U)$. Il est appelé le "Hom interne" ou encore le faisceau des germes de morphismes de A vers B.

Il est trivial de vérifier que le foncteur $(\bullet)^A$ est *adjoint à droite* du foncteur $A \times (\bullet)$. Cela signifie que pour tout objet C de $\mathfrak{F}$ il existe un isomorphisme fonctoriel $\mathrm{Hom}_{\mathfrak{F}}(C, B^A) \cong \mathrm{Hom}_{\mathfrak{F}}(A \times C, B)$. Par exemple, pour $C = 1$, on obtient $\mathrm{Hom}_{\mathfrak{F}}(1, B^A) \cong \mathrm{Hom}_{\mathfrak{F}}(A, B)$. Mais un morphisme $f: 1 \rightarrow B^A$ est comme un "élément" de B^A. En fait, si A est un faisceau, un morphisme $s : 1 \rightarrow A$ est une section globale de A, c'est-à-dire un élément $s \in A(M)$.

L'isomorphisme $\mathrm{Hom}_{\mathfrak{F}}(C, B^A) \cong \mathrm{Hom}_{\mathfrak{F}}(A \times C, B)$ est évident dans le topos des ensembles: si f_c est une famille d'applications $f_c : A \rightarrow B$ alors elle est équivalente à l'application $f: A \times C \rightarrow B$ définie par $f(a, c) = f_c(a)$.

Si l'on considère $C = B^A$ et Id_{B^A}, l'adjonction à droite définit ce que l'on appelle une *co-unité* $\varepsilon : A \times B^A \rightarrow B$ telle que pour chaque $f: C \rightarrow B^A$ (et donc $1 \times f: A \times C \rightarrow A \times B^A$) le morphisme associé $h : A \times C \rightarrow B$ soit donné par $h = \varepsilon \circ (1 \times f)$. La co-unité généralise l'application *d'évaluation* $(x, f) \rightarrow f(x)$ de la théorie des ensembles.

La catégorie $\mathfrak{F}$ possède en outre un *classificateur de sous-objets,* c'est-à-dire un monomorphisme *True*: $1 \rightarrow \Omega$ tel que tout sous-objet S d'un objet A (i.e. tout monomorphisme $m : S \rightarrow A$) soit reconstructible par pull-back à partir d'une "fonction caractéristique" $\varphi_S : A \rightarrow \Omega$ de S :

$$\begin{array}{ccc} S & \rightarrow & 1 \\ \downarrow & & \downarrow \mathrm{True} \\ A & \underset{\varphi_S}{\rightarrow} & \Omega \end{array}$$

Ω est l'ensemble des "valeurs de vérité" du topos $\mathfrak{F}$. Ce classificateur "internalise" donc les sous-objets. Il en représente le foncteur. Dans la catégorie des ensembles **Ens**, $\Omega = \{0, 1\}$ correspond aux valeurs de vérité booléennes. C'est le fait que Ω puisse être beaucoup plus compliqué qui fait l'intérêt majeur de la théorie des topoi. Nous allons voir que c'est ce qui permet dans notre cas de *localiser la vérité* conformément à l'idée de Thom.

A partir de ces constructions catégoriques, on peut définir d'autres constructions et donner un sens aux concepts "ensemblistes" d'élément, de propriété et de partie.

1. Par exemple, un morphisme $\vartheta : A \to \Omega$ est un "prédicat" de A, c'est-à-dire une propriété de ses "éléments généralisés" $a : B \to A$. On a $\vartheta(a)$ ssi a "appartient" au sous-objet S de A défini par ϑ ($\varphi_S = \vartheta$), i.e. ssi $\varphi_S \circ a = True_B$ avec $True_B : B \to 1 \to \Omega$.
2. De même, les parties de A (ses sous-objets) sont définies par $P(A) = \Omega^A$. Dans la catégorie des ensembles **Ens**, $\Omega = \{0, 1\}$ correspond aux valeurs de vérité booléennes. Il en va tout autrement ici. Le faisceau Ω est défini par $\Omega(U) := \{V \subset U\}$ et $True : 1 \to \Omega$ par $True(U) : 1 \to U \in \Omega(U)$, i.e. par l'élément *maximal* de $\Omega(U)$. Cela signifie que:
   (i) la vérité devient *spatialement localisée*,
   (ii) "être vrai sur U" signifie "être vrai partout sur U".

$\Omega(U)$ n'est pas une algèbre de Boole (car le complémentaire d'un ouvert n'est pas un ouvert mais un fermé). C'est une algèbre de Heyting. Ω est donc un faisceau d'algèbres de Heyting.

6.6. Topoï et logique

Le point fondamental est que l'on peut canoniquement associer à un topos $\mathfrak{F}$ une *logique interne,* c'est-à-dire:

(i) un *langage formel* $L_{\mathfrak{F}}$ appelé son langage de Mitchell-Bénabou,

(ii) une *sémantique* de type forcing appelée sa sémantique de Kripke-Joyal.

Comme nous l'avons vu, l'idée directrice est qu'un faisceau $X \in \mathfrak{F}$ peut être traité comme un *type* pour des variables x qui seront interprétées comme des *sections* s de X, $s \in X(U)$. On obtient donc *à la fois une typification et une localisation spatiale* des variables et c'est exactement ce dont nous avons besoin pour relier la géométrie des sections à une logique du jugement: *la syntaxe des jugements porte sur les types et la sémantique sur la localisation.* La sémantique concerne donc (dans le cas qui nous occupe ici) les conditions de vérité associées aux phénomènes de recouvrement de domaines spatiaux par des qualités. Ce qui est précisement la façon Thom définit la vérité des jugements perceptifs.

6.6.1. Syntaxe

De façon générale, la structure de topos d'une catégorie $\mathfrak{F}$ permet de construire récursivement les termes σ du langage $L_{\mathfrak{F}}$ comme des *morphismes* entre faisceaux.

Le langage possède des *types*: 1, Ω (type des formules ou valeurs de vérité), des types de base, des types produits d'autres types ΠX_i, un type $P(X)$ de parties pour chaque type X, des variables typées $x : X$ pour chaque type X et des symboles de fonction $f : A \rightarrow B$ entre types. On définit alors récursivement les *termes* typés de façon standard.

1. Il existe un terme unique $* : 1$ de type 1.
2. Si $\tau : A$ [12] et $f : A \rightarrow B$ alors il existe un terme image $f(\tau)$.
3. Si $\tau_i : A_i$ alors il existe un terme produit $\tau = < \tau_i > : \Pi A_i$ et réciproquement si $\tau : \Pi A_i$ on sait définir ses projections $\tau_i : A_i$.
4. Si $\varphi(a) : \Omega$ est une formule avec $a : A$ alors il existe un terme "partie de A satisfaisant φ" $\{a \mid \varphi\} : P(A)$.
5. Si $\sigma, \tau : A$ alors on peut construire le terme $\sigma = \tau : A$.
6. Si $\sigma : A$ et $\tau : P(A)$ alors on peut construire le terme $\sigma \in \tau : \Omega$.

L'interprétation dans un topos se fait alors de la façon suivante.

$* : 1$ est interprété par l'objet terminal 1. Si le terme σ est de type A et est construit à partir de variables libres x_i de types respectifs X_i, il est interprété par un morphisme $\sigma : \Pi X_i \rightarrow A$ qui exprime sa structure. Le fait décisif est que la structure de topos est précisément celle qui permet de définir toutes les structures logiques nécessaires en prenant Ω *comme type pour les formules*.

On définit par exemple les termes $< \sigma, \tau >$ (paire ordonnée), $\sigma = \tau$ (égalité), $f \circ \sigma$ (termes composés), $\phi(\sigma)$ (termes fonctionnels), $\sigma \in \tau$ (apparte-

12 La notation $\tau : A$ signifie que le terme τ est de type A.

nance), $\lambda x\sigma$ (λ-termes), ainsi que les quantificateurs. Ces derniers sont les adjoints à droite et à gauche des morphismes "image inverse" $P(f) : P(B) \to P(A)$ (où $P(A) = \Omega^A$ est le faisceau des "parties" de A) canoniquement associés aux morphismes $f : A \to B$.

Donnons quelques exemples.

1. Soient $\sigma : X \to A$, $\tau : Y \to A$ deux termes de même type A. Ils permettent de définir le terme $\sigma = \tau$ de type Ω interprété par:
$$\sigma = \tau : Z = X \times Y \to A \times A \xrightarrow{\delta_A} \Omega$$
où δ_A est la fonction caractéristique du sous-objet diagonal $\Delta : A \to A \times A$. C'est la diagonale qui traduit l'égalité. En théorie des ensembles σ et τ correspondent à des éléments a et b de A et la diagonale correspond aux paires (a, b) telles que $a = b$.

2. Des termes $\sigma : X \to A$ et $\tau : Y \to \Omega^A$ définissent de même un terme $\sigma \in \tau$ de type Ω interprété par :
$$\sigma \in \tau : Z = X \times Y \to A \times \Omega^A \xrightarrow{e} \Omega$$
où e est le morphisme d'évaluation. En théorie des ensembles, σ correspond à un élément a de A et τ à la fonction caractéristique d'un sous-ensemble de A.

6.6.2. Sémantique

Quant à la sémantique de Kripke-Joyal, c'est une sémantique de type *forcing*. Elle repose sur des règles du type $U \Vdash \varphi(s)$ où U est un ouvert de M, s une section de $X(U)$, x une variable de type X et $\varphi(x)$ une formule. Quant au langage (syntaxe) x est une variable typifiée, mais quant à sa dénotation (sémantique) elle est une section s i.e. un remplissement qualitatif d'une extension spatiale. La relation de forcing $\Vdash$ signifie *que l'extension U localisant la vérité dans U valide $\varphi(s)$*. Autrement dit, si $x : X$, si $\varphi(x)$ est une formule et si $s \in X(U)$, alors $U \Vdash \varphi(s)$ ssi $s \in \{x \mid \varphi\}(U)$.

Les règles sémantiques naturelles pour les connecteurs logiques, l'implication, la négation et la quantification montrent que la logique interne d'un topos est en général de nature *intuitionniste*. Elles sont données par :

(i) $U \Vdash \varphi(s) \wedge \psi(s)$ ssi $U \Vdash \varphi(s)$ et $U \Vdash \psi(s)$. Cette règle est classique et dit que $\varphi(s)$ et $\psi(s)$ doivent être toutes deux globalement réalisées sur U.

(ii) $U \Vdash \varphi(s) \vee \psi(s)$ ssi il existe un recouvrement ouvert $\mathfrak{U} = (U_i)_{i \in I}$ de U tel que pour tout i on ait $U_i \Vdash \varphi(s|_{U_i})$ ou $U_i \Vdash \psi(s|_{U_i})$. Cette règle pour la disjonction est intuitionniste. Elle dit que, localement sur U, au moins l'une des $\varphi(s)$ ou $\psi(s)$ doit être réalisée. Si V (resp. W) est l'union des U_i forçant $\varphi(s|_{U_i})$ (resp. $\psi(s|_{U_i})$), alors $U = V \cup W$ avec $V \Vdash \varphi$ et $W \Vdash \varphi$.

(iii) $U \Vdash \varphi(s) \Rightarrow \psi(s)$ ssi, pour tout $V \subseteq U$, $V \Vdash \varphi(s|_V)$ implique $V \Vdash \psi(s|_V)$.

(iv) $U \Vdash \neg\, \varphi(s)$ ssi il n'existe pas de $V \subseteq U$, $V \neq \varnothing$ tel que $V \Vdash \varphi(s|_V)$ (la négation est intuitionniste car Ω est une algèbre de Heyting et non pas une algèbre de Boole).

(v) $U \Vdash \exists y\ \varphi(s, y)$ (y étant de type Y) ssi il existe un recouvrement ouvert $\mathcal{U} = (U_i)_{i \in I}$ de U et des sections $\beta_i \in Y(U_i)$ tels que pour tout $i \in I$ on ait $U_i \Vdash \varphi(s|_{U_i}, \beta_i)$.

(vi) $U \Vdash \forall y\ \varphi(s, y)$ ssi pour tout $V \subseteq U$ et $\beta \in Y(V)$ on a $V \Vdash \varphi(s|_V, \beta)$.

Les subtilités de cette forme de sémantique résultent du problème suivant. Dans un topos, on peut interpréter des expressions de type ensembliste comme:

$$\left\{ (x_i)_{i \in I} \in \prod_{i \in I} X_i \,\middle|\, \varphi(x_i) \right\}$$

définies par $(a_i) \in \{(x_i) \mid \varphi(x_i)\}(U)$ ssi $U \Vdash \varphi(a_i)$. Mais on doit s'assurer que de telles expressions définissent des sous-faisceaux. Pour cela, il faut "faisceautiser" les sous-foncteurs intervenant dans de telles constructions.

7. La structure des jugements perceptifs: combler le hiatus morphologique/logique

Revenons à notre problème thomien-husserlien des liens entre la présentation perceptive <S, p> et la représentation prédicative "S est p". Soient M l'espace et C la fibration (triviale) de base M et de fibre l'espace G des couleurs. Le genre qualitatif G est catégorisé en espèces $C_1, C_2, \ldots, C_n$. Soit $\mathfrak{C}$ le faisceau des sections de C. Aux différentes catégories (au sens banal, non mathématique) C_i, correspondent des sous-faisceaux $\mathfrak{C}_i$ de $\mathfrak{C}$.

Soit S un symbole dénotant un individu. S est un *index* qui, dans chaque monde possible $\mathfrak{M}$, *sélectionne* son extension $W_S \in \mathfrak{O}(M)$ et, pour tout genre pertinent de qualité G, une section $g_S \in \mathfrak{C}(W_S)$. Cela signifie que, relativement à G, S s'interprète comme un morphisme $g_S : W_S \rightarrow \mathfrak{C}$, c'est-à-dire *comme un point de* $\mathfrak{C}$ *à valeurs dans* W_S. Soit alors p la catégorie C_i considérée et $\mathfrak{p}$ le *sous-faisceau* du faisceau $\mathfrak{C}$ qui lui est associé. On peut retraduire l'interprétation thomienne (ante-prédicative, pré-judicative) du jugement "S est p" en disant:

"S est p" est vrai ssi $g_S \in \mathfrak{p}(W_S)$.

Le sous-objet $\mathfrak{p}$ de $\mathfrak{C}$ correspond à un prédicat sur $\mathfrak{C}$, c'est-à-dire à un morphisme $\varphi_{\mathfrak{p}} : \mathfrak{C} \rightarrow \Omega$. $\varphi_{\mathfrak{p}}$ associe à toute section g de $\mathfrak{C}(U)$ l'ouvert maximal V de U sur lequel g est à valeurs dans p. On a alors:

"S est p" est vrai ssi $\varphi_{\mathfrak{p}}(g_S) = \text{True}$.

On retrouve ainsi l'interprétation ensembliste-prédicative (tarskienne) standard:

"S est p" est vrai ssi $p(S) = \text{True}$,

mais en ayant tenu compte de l'extension spatiale de l'entité S considérée pour localiser la vérité.

8. Le problème des bords d'objets

Nous avons signalé plus haut la difficulté concernant le fait que l'extension des objets perçus est en général *fermée* dans la mesure où les objets sont limités par des bords. Qui plus est, on sait depuis Brentano que les bords ont un statut quelque peu paradoxal. Il y a plusieurs façons de prendre en compte cette difficulté. Nous en évoquerons deux pour conclure.

8.1. *Les co-algèbres de Heyting*

Une première idée, due à Lawvere, est d'utiliser des co-algèbres de Heyting. Dans une algèbre de Heyting d'ouverts, la négation $\neg U$ de U est l'intérieur de son fermé complémentaire, c'est-à-dire le plus grand ouvert V tel que $U \cap V = \varnothing$. Dualement, dans une co-algèbre de Heyting de fermés, la négation $\neg F$ de F est la fermeture de son ouvert complémentaire, c'est-à-dire le plus petit fermé H tel que $F \cup H = 1$. On a évidemment $\neg(F \cap H) = \neg F \cup \neg H$, mais seulement $\neg(F \cup H) \subseteq \neg F \cap \neg G$.

On définit alors le bord ∂F de F comme l'intersection $F \cap \neg F = \partial F$. ∂F est donc défini en termes de contradiction logique.

L'opérateur bord satisfait la règle de Leibniz:

$$\partial(F \cap H) = (\partial F \cap H) \cup (F \cap \partial H).$$

Les bords B sont caractérisés par $\partial B = B$ c'est-à-dire par $\neg B = 1$ ou $\neg\neg B = 0$. De façon générale, la double négation $\neg\neg F \subset F$ (la fermeture de son intérieur) est le noyau régulier (le "regular core") de F. On a bien sûr $F = (\neg\neg F) \cup \partial F$.

8.2. Les fibrations stratifiées

Une autre façon d'introduire les bords dans le formalisme toposique est, de façon très générale, de tenir compte du fait que les remplissements intuitifs d'extensions par des qualités sont *segmentés par des discontinuités qualitatives K.*[13] Les sections de fibration g exprimant ces remplissements sont discontinues le long de K. Par ailleurs il est justifié de faire l'hypothèse que K est un ensemble fini de morceaux de courbes régulières C_i s'arrêtant en des points d'arrêt A_j et se connectant à travers des points multiples T_k (génériquement des points triples) comme des jonctions en T. En effet les meilleurs modèles de segmentation dont on dispose (en particulier le modèle variationnel dit de Mumford-Shah) semblent posséder cette propriété (la conjecture est presque démontrée).

Cela signifie que *K stratifie M.* Les strates de dimension 2 sont les composantes connexes du complémentaire de K. Elles sont ouvertes. Ce sont les régions homogènes de remplissement qualitatif. Les strates de dimension 1 sont les morceaux de courbes C_i et les strates de dimension 0 les points singuliers isolés A_j et T_k.

On peut alors reprendre le formalisme faisceautique du recollement de sections locales mais

(i) en considérant des ouverts stratifiés (U, K_U) et en imposant des règles de compatibilité dimensionnelle pour le recollement des strates, et

---

13 La segmentation est un problème fondamental posé par Husserl à la suite de Stumpf (le fondateur de la Gestalttheorie). Il a connu ces dernières années un approfondissement considérable. Cf. Petitot (2003b).

(ii) en permettant aux sections d'être discontinues le long des strates singulières de K_U.

Qui plus est, pour que le formalisme proposé soit plausible, il faudrait également étendre les formulations "faisceautiques" et "toposiques" aux cas où les fibres des fibrations considérées sont *catégorisées* (au sens cognitif). La catégorisation cognitive pose elle aussi un problème fondamental pour lequel il existe des modèles morphodynamiques.

9. Conclusion

Nous pensons que l'approche proposée ici constitue le formalisme de complexité minimale permettant, et uniquement dans le cas des propositions les plus triviales, de faire explicitement le lien entre perception et prédication en partant des conditions de vérité définies par René Thom. Ces développements formels prouvent qu'il est possible d'articuler le schématisme morphologique sur une sémantique formelle de nature *indexicale* (donc en fait plus pragmatique que proprement sémantique). Dans une telle approche, il apparaît que l'espace fonctionne bien comme une *modalité* dans la mesure où la sémantique des situations où les variables sont à la fois typifiées et localisées est une sémantique modale à la Kripke. Cela permet de réinterpréter le caractère "synthétique a priori" de l'espace.

On voit à quel point les approches cognitives obligent à reproblématiser ce qui avait été trivialisé par l'analytique logique classique. Si on ajoute à cela les problèmes d'implémentation neuronale, on voit que les conditions d'application et de vérité du plus simple des jugements perceptifs se situe à un niveau de profondeur sans aucune mesure avec la tautologie tarskienne et que "le hiatus entre le logique et le morphologique" que René Thom considérait comme "infranchissable" n'est peut-être pas vraiment infranchissable mais est certainement très dur à franchir.

Bibliographie

Asperti, A. (Longo). G., 1991. *Categories, Types, and Stuctures*, MIT Press, Cambridge.

Husserl, E. (1954). *Erfahrung und Urteil, Untersuchungen zur Genealogie der Logik*, Claassen & Goverts, Hamburg.

Mac Lane, S., Moerdijk, I. (1992). *Sheaves in Geometry and Logic*, Springer, Berlin.

Moerdijk, I., Reyes, G. (1991). *Models for Smooth Infinitesimal Analysis*, Springer, Berlin.

Petitot, J. (1992). *Physique du Sens,* Editions du CNRS, Paris.

Petitot, J. (1994a). "Phenomenology of Perception, Qualitative Physics and Sheaf Mereology", *Philosophy and the Cognitive Sciences*, Proceedings of the 16th International Wittgenstein Symposium (R. Casati, B. Smith, G. White eds), Verlag Hölder-Pichler-Tempsky, Vienna, 387–408.

Petitot, J. (1994b). "La sémiophysique: de la physique qualitative aux sciences cognitives", *Passion des Formes*, à René Thom (M. Porte éd.), 499–545, E.N.S. Editions, Fontenay-Saint Cloud.

Petitot, J. (1995). "Sheaf Mereology and Husserl's Morphological Ontology", *International Journal of Human-Computer Studies*, 43, 741–763.

Petitot, J. (2003a). "The neurogeometry of pinwheels as a sub-riemannian contact structure", *Neurogeometry and Visual Perception* (J. Petitot, J. Lorenceau eds), *Journal of Physiology-Paris*, 97, 2–3, 265–309.

Petitot, J. (2003b). "An introduction to the Mumford-Shah segmentation model", *Neurogeometry and Visual Perception* (J. Petitot, J. Lorenceau eds), *Journal of Physiology-Paris*, 97, 2–3, 335–342.

Thom, R. (1988). *Esquisse d'une Sémiophysique*, InterEditions, Paris.

René Thom – Prégnances et catastrophes. Pour une phénoménologie sémio-cognitive

PER AAGE BRANDT

> Une chose me semble assez évidente: de même que la conscience chez l'homme est liée [...] à la conscience de l'espace et du temps, [à] la représentation interne d'un cadre spatio-temporel contenant notre corps: de la même manière, on peut supposer que les animaux supérieurs disposent, eux aussi, "psychiquement", d'une carte représentant leur territoire et contenant leurs corps, leur organisme comme sous-ensemble particulièrement distingué. C'est à partir de cette idée de base que je vais essayer de vous expliquer les notions de saillance et de prégnance. Mon *a priori*, c'est essentiellement la conscience que nous avons de l'espace et du temps, de l'espace qui nous entoure et du temps.
>
> "Saillance et prégnance", 1991[1] (Actes d'un colloque UNESCO de 1988).

1. Les deux théories sémiotiques de René Thom

Comme on sait, le mathématicien spécialisé en topologie que fut René Thom avait, dans le domaine philosophico-sémiotique qui l'intéressait de plus en plus dans la seconde partie de sa carrière, développé l'essentiel de deux théories qui ont toutes deux fait l'objet de commentaires et de reprises par bien d'autres chercheurs dans ce domaine: la *théorie des catastrophes* et la *théorie des saillances et des prégnances*. La première propose des modèles topologiques permettant, entre beaucoup d'autres choses biologiques, de représenter et de décrire certaines structures dynamiques d'articulation des strates de valeurs et des actants dans la sémantique du vécu phénoménal, ces formations locales que nous appelons parfois aussi des schématismes cognitifs; ces représentations permettent de modéliser des conflits entre forces se déployant dans l'espace et le temps de l'expérience. Dans cette modélisation

1 La publication, en 2003, des Oeuvres Complètes de René Thom sur CD-ROM par l'Institut des Hautes Etudes Scientifiques, à Paris, a grandement facilité l'étude du mathématicien-philosophe-sémioticien.

topologique, il s'agit de voir les états et les transitions entre états de deux manières complémentaires:

1. Un système de référence S est situé dans un paysage "réel" R d'états possibles, et la stabilité de chaque état dépend de l'ensemble d'états possibles dont il fait partie, dans R, dans la mesure où un état pour S est un attracteur, et que la *co-présence des attracteurs* crée la possibilité pour S de parcourir ce paysage topologique comme un ensemble d'itinéraires possibles de changements d'états; c'est là l'*espace interne* du modèle catastrophiste.

2. Comme le paysage R est défini par un potentiel à variables internes, de structure, (x, y,...) et à variables externes, de contrôle, (a, b,...), par exemple celui, toujours assez simple, de la catastrophe cusp: $y = x^4 + ax^2 + bx$, les propriétés topographiques de l'espace interne (x, y,...), c'est-à-dire le nombre et la puissance relative des attracteurs, sont *déterminées par l'espace externe* C des variables de contrôle (a, b,...).

Le temps interne est celui des changements du système de référence S le long de son trajet pour ainsi dire narratif. Le temps externe est celui, pour ainsi dire géologique, des transformations du paysage qui conditionne le trajet de S.

Le rapport de forces dans une topologie dynamique R varie donc selon la magnitude relative et modifiable des attracteurs et des répulseurs, déterminée dans C; certaines variations dans C gardent la probabilité des trajets dans R, alors que d'autres, critiques, la changent. Ainsi, un même modèle, biplane (R, C), rend compte des continuités (états) et des discontinuités (événements), des ruptures de continuité critiques, de ce qui est "même" et de ce qui est "autre", dans les circonstances R de S.[2] Les espaces de contrôle sont à la fois mathématiquement simples et sémantiquement interprétables, si la topologie fait partie de l'ensemble des "catastrophes élémentaires", et surtout de

2 Ces topologies critiques sont d'une grande efficacité quand il s'agit, en sémiolinguistique, de décrire des transformations modales – par exemple, des situations où un état de choses qui était possible devient impossible ou, au contraire, nécessaire. Ou de la différence entre une réalisation normale (possible, probable ou nécessaire) et une réalisation événementielle miraculeuse ou catastrophique (impossible). Les valeurs modales sont de très purs cas de schématismes dynamiques catastrophiquement modelables. Cf. P. Aa. Brandt (1993).

l'ensemble des "cuspoïdes".[3] Cet ensemble pourrait ainsi constituer un inventaire élémentaire des formes paradigmatiques ou syntagmatiques du sens vécu ou même imaginable (dans l'espace-temps 3D).[4]

Or, l'articulation de ce premier volet de la *sémantique spatio-temporelle* de Thom et de son second volet, celui dont il s'agira maintenant, reste à établir; cette articulation constitue même la difficulté centrale de la sémiotique thomienne. Nous verrons néanmoins comment la première modélisation fonde une *sémiotique dynamique* comme pensée du *signe*, et comment celle-ci sera interprétée dans le cadre de la seconde modélisation.

Dans la seconde théorie, en effet, il s'agit de voir de quelle manière un objet d'attention, une *saillance*, signifie, par la "prégnance" qui l'investit. Les saillances sont des événements perceptifs, précisément des ruptures de continuité dans le sensible, susceptibles d'attirer l'attention d'un organisme conscient – tels un cri dans le silence, une tour qui s'élève sur le village, une aspérité sur une surface lisse – et qui invitent un tel être à les interpréter dans des termes organiquement appropriés: ce qui rend significatif, "prégnant" de sens, un objet ou un événement saillant, c'est ce par quoi – comme l'odeur de quelque chose de comestible, de sexuel, de dangereux, d'"intéressant" en général – l'entité saillante attire l'attention, non seulement vers elle-même, mais donc, par là, vers quelque chose d'autre qui se situe "derrière" ou se trouve proche ou autrement lié à ce premier attracteur dans l'espace ou le temps. L'attention, interprétée comme système S, se fixe, puis se déplace, glisse, court ou vole du premier attracteur, *saillant*, vers un second attracteur, source de la *prégnance*. Phénoménologiquement parlant, la double attraction de la saillance puis de la prégnance venant comme un flux d'une source cachée fait de l'attention un nageur à contre-courant qui suit en amont ce flux de la prégnance qui, elle, afflue en aval vers l'entité saillante. Thom écrit, au sujet du flux prégnanciel dans un mode bio-sémiotique:

> Une prégnance biologique se comporte donc comme un *fluide érosif* qui s'infiltre à travers ces fractures du réel que constituent les formes saillantes. Thom (1990: 460) (italiques par l'auteur).

---

3 C'est-à-dire le déploiement des germes: x^3 (le pli), x^4 (le cusp), x^5 (la queue d'aronde), x^6 (le papillon).

4 Le sens vécu se compose, dans notre optique, d'événements internes, épisodiques, actantiels: S passe de l'état X à l'état Y, et d'événements externes, circonstantiels, géologiques, ou écologiques, si l'on veut, concernant la qualité dynamique de l'espace-temps où se déroule le fait épisodique.

Les besoins physiologiques, la faim et la soif, le besoin de protection, le désir sexuel, sont des exemples évidents de telles *prégnances* prédominantes dans le règne animal. Ainsi:

> [...] il faut concevoir que la *prégnance alimentaire* se comporte, dans le champ phénoménal des formes vécues, comme un *fluide invasif* qui se propage à travers les fissures du champ que constituent les formes saillantes. (Thom, 1991) (italiques par l'auteur)

Et en général, odeur ou lumière, la prégnance imprègne:

> [...] On peut considérer la prégnance comme une sorte de source attachée à certaines formes qui *illuminent* certaines structures constituées par les formes saillantes [...] (ibidem) (italiques par l'auteur).

Thom reprend l'exemple bien connu du chien de Pavlov qui perçoit (interprète) le bruit de la sonnette comme un déclencheur d'attente de l'arrivée devant lui d'un morceau de viande. L'attention suit la trace qui mène du terme de la prégnance auprès de la saillance, et jusqu'à sa source cachée, mais proche:

At (attention) → Sa (saillance: sonnette) → Pr (prégnance: viande)

Le son entendu par le chien est, pour ainsi dire, perçu par lui comme investi d'une odeur mentale de viande. C'est ce qui rend “intéressante” la sonnette pour le chien; c'est ce qui “signifie” la viande et permet d'en attendre la présentation imminente. On pourrait parler d'une olfactivité mentale, et en ce sens, d'une sémiotique du <u>sentir</u>...

Il semble encore s'agir d'une bio-sémiotique assez primaire; mais Thom entend bien viser en dernière instance le sens vécu dans le monde humain et manifesté dans le langage:

> [...] Quand je parle de prégnances du langage humain, j'envisage une équivalence de type “prégnance = *concept*”. Le concept, au sens philosophique du terme, a un aspect d'extension et un aspect d'intension; l'extension représente la saillance; l'intension, la signification, le sens, représente[nt] la prégnance.”
>
> Toute cette problématique de l'investissement d'une forme saillante par une prégnance et des effets figuratifs qui s'ensuivent peut être considérée, à mon sens, comme l'objet d'une *sémiotique généralisée* [...] (Thom 1991) (italiques par l'auteur).

Un mot, un terme, est ainsi une saillance, qui se donne chaque fois que ce terme est appliqué à quelque chose; c'est l'extension de son concept. A chaque fois, le sens du terme, sa prégnance, est senti comme un parfum mental qui l'enveloppe, l'embaume, et qui renvoie l'animal destinataire (S) à ce que le mot signifie (intension conceptuelle, dans un paysage conceptuel R); alors que le flux venant de cette source et se répandant dans le sensible en constitue l'extension. C'est là une conception apparemment assez vague et impressionniste, mais au fond, d'une certaine efficacité: le *signe* serait ainsi le support présent, saillant, prégnant d'un "halo" sémantique manifestant un sens invisible, mais proche. On peut, si l'on veut, ainsi figurativement représenter le rapport saussurien du signifiant au signifié dans les signes, tels qu'ils existent et fonctionnent dans l'expérience humaine: il s'agirait d'un rapport entre saillance (substance de l'expression[5]) et prégnance (forme de l'expression, ou *signifiant*, comme promesse, c'est-à-dire version formelle, *signifée*, d'un contenu substantiel). La prégnance afflue et se manifeste comme parfum signifiant d'une saillance; la lecture du signe consiste à suivre le fil olfactif jusqu'à sa source substantielle. Il s'agirait d'une sorte de bio-phénoménologie de la fonction sémiotique.

2. Les signes et le double flux

Essayons de suivre et de développer cette idée. Un signe naturel, un *indice*, dans la terminologie de C. S. Peirce, est typiquement (dans le temps) un *effet* qui envoie l'attention du sujet vers l'idée de la cause de cet effet; ou bien (dans l'espace) une *partie* qui renvoie de la même manière au tout dont elle est une partie. La signification, par exemple, de la fumée (perçue) est l'attente du feu (non perçu). C'est-à-dire, dans cette conception, que la *source de la prégnance*, le référent réel feu s'est fait connaître par une *saillance prégnante*, donc signifiante, /fumée/, qui serait liée par une fonction de signe à la signification, ou le signifié, /feu/. Il y a d'abord un flux qui va du référent substantiel (ici: la cause) vers la prégnance, le signifiant (ici: l'effet); ensuite il y a une lecture en sens inverse qui va du signifiant vers le signifié (ici: l'idée de la cause). Dans l'imaginaire perceptif du sujet, organisme conscient, éventuellement humain, le sens naturel se donne, *vient vers lui*

5 Louis Hjelmslev (1943) distingue pertinemment dans la fonction sémiotique (plan de l'expression, plan du contenu) entre substance et forme de l'expression, et entre substance et forme du contenu.

(puisqu'il part du référent pour investir quelque chose en signifiant), et l'être substantiel de ce sens se cache, existe comme signifié pour le sujet qui *va vers lui*, à partir de l'instance saillante qu'il investit, et qui donc le signifie. Résumons cette logique fluide du signe dans le cas simple de l'indice:

Source de prégnance ————flux————> saillance investie: sa
(cause; tout)<————sens—————prégnance de sé (effet; partie)
[substance du contenu] [forme ou idée du contenu]

A gauche: le référent, source et but de la boucle sémiotique; à droite: l'interprétation organique et subjective de la saillance comme prégnance, c'est-à-dire comme "signifiance", *sa*, indication de la proximité de quelque chose dont l'interprète peut se donner une idée (*sé*), signifiée par cette prégnance.

Dans le cas de la clochette et du chien, la prégnance première, l'odeur réelle de la viande, est transportée par le conditionnement, et devient à son tour source de prégnance; par un curieux principe de transitivité cognitive, le son de la clochette pavlovienne signifiera donc /viande/ pour le chien. Nous pouvons remarquer que l'association entre son de clochette et présentation de viande doit être liée, pour le chien, à l'intervention d'un être humain, l'expérimentateur, ou du moins à la stabilité d'un scénario. (Et le chien ne croit pas que la clochette elle-même se mange, par un effet de transsubstantiation!). L'esprit animal est déjà prêt à effectuer une certaine forme, élémentaire, de symbolisation. Le transfert de prégnance ne dépend pas d'une perception "causaliste" directe; il peut, comme chez les humains, se satisfaire d'un lien "magique", ou arbitraire.

C'est cette logique de la prégnance qui nous permettra d'établir la relation entre les deux motifs théoriques de Thom: la figurativité topologique formalisera, rendra formelle, à la fois la composante "substantielle" du flux prégnanciel et la composante "formelle" de son interprétation par un sujet qui inversera ce flux et qui part mentalement, cognitivement, puis physiquement, à la recherche de sa source.

La représentation d'un attracteur, dans la topologie des potentiels différentiables dont il s'agit, est le minimum d'une courbe de la fonction $y = f(x)$, tel le minimum à (0, 0) de la parabole ($y = px^2$). Un système énergétique dépense plus d'énergie en sortant d'un état qu'en y restant; cette intuition sert de principe "gravitationnel" pour la convention qui veut que les maxima se lisent comme des répulseurs et les minima comme des attracteurs, pour un système S.

Puisqu'un minimum, un puits de potentiel, est donc lu comme un attracteur dynamique, voici donc, dans l'indice, un puits référentiel RE (la source prégnantielle) et un puits SA (la saillance investie de prégnance). Le flux de ce qui se manifeste et s' "exprime" est donc un système réel qui va de l'état RE à l'état SA. On pourra appeler ce premier flux *le sens*. Le sens se propage réellement (dans la réalité du sujet) de RE à SA, puis éventuellement à d'autres SA^{n+1}, dans la mesure où de tels attracteurs existent, ou surgissent (crées par les opérations ayant lieu dans l'espace de contrôle).

Or, la lecture de ce *sens*, par le sujet, consiste à suivre le même flux, celui du sens, à contre-courant, vers sa source. C'est l'attention du sujet qui va de l'attracteur saillant à la recherche d'un attracteur sous-jacent, dans la mesure où SA est interprété comme le signifiant d'un sens signifié /RE/, conceptuellement figuré ou indéterminé, vague ou définissable, selon le cas. Le sujet part cognitivement d'un SA^p (marqué par une prégnance) vers ce qui "fait sens" et qui est anticipé comme /RE/.

Cette manière plus ou moins bio-sémiotique de concevoir la fonction sémiotique et l'expérience du signe correspond donc à deux variations sur une topologie de cusp, distinguées par la force relative des attracteurs RE, SA (et RE^p, /SA/).

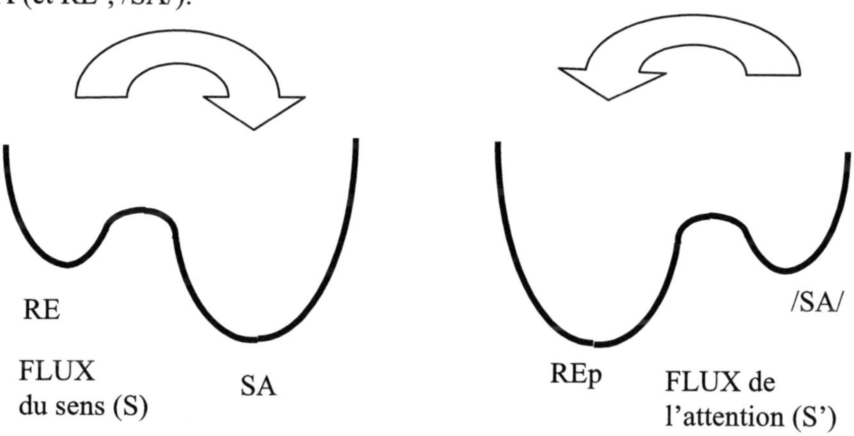

Figure 1. Flux du sens et flux de l'attention.

Pour le *sujet*, en effet, c'est-à-dire le chien, l'humain, ou en général le "système" conscient dont on décrit la situation sémiotique, le sens découle *objectivement* d'une source cachée et vient s'investir dans une entité saillante; pour *l'observateur*, en revanche, le sens est bien sûr une projection *subjective* venant de ce sujet, de sa perception des choses comme signifiantes, de sa lecture du monde qui projette son attention vers quelque chose de sous-jacent, quelque chose qui serait là comme indiqué, signifié, par un objet saillant. Flux primaire du sens lui-même, à partir du monde, ou d'un destinateur; flux secondaire de l'attention intentionnelle du destinataire, allant à son encontre – les deux sont essentiels dans la perspective d'une sémiotique réaliste. Dans le réel du sujet, dans son paysage R, le sens se cache et se donne à connaître de cette manière. Or on voit bien que le rapport de forces entre les attracteurs doit s'inverser dans la réponse sémiotique du sujet au défi de la prégnance. Le paysage R est modifié entre les deux phases, ou les deux flux (RE $\rightarrow$ SA) et ($SA^p \rightarrow$ /RE/).

Techniquement, l'espace de contrôle C, qui – il faut bien le comprendre – fait partie de la phénoménologie dynamique du sujet au même titre que l'espace des attracteurs, effectue une variation de p à q; dans la perspective de C, la dynamique sémiotique se présente comme suit (Fig. 2):

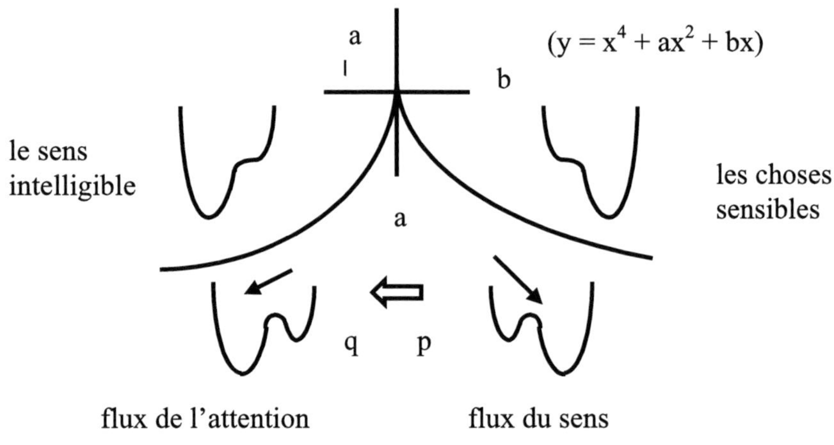

Figure 2. Dynamique sémiotique.

Si une prégnance est suivie d'une autre prégnance, on aura une répétition du phénomène, une bifurcation de la prégnance, toujours selon la même logique topologique. Pour que quelque chose fasse sens, il faut d'abord que le sens afflue vers le monde sensible; ce qui demande une conjoncture p. C'est là, pour le sujet, déjà un problème ontologique: *d'où vient* la bifurcation entre le puits RE et le puits SA? *Cognitivement*, le sujet sent que cela "veut" ou "doit"se produire, puisque l'avènement de cet événement se doit à une opération dans l'espace C (vécu comme "le destin", "le narrateur", voire "le ciel" ...), c'est-à-dire que la manifestation du sens, même de le cas des indices, est en quelque sorte intentionelle, “voulue” ou “due” quelque part. L’inversion p → q, également due à une opération dans C, est cognitivement sentie par le sujet comme sa réponse naturelle et spontanée à l’appel du sens – on peut parler d’un schématisme “sémiophilique” inhérent à la prégnance. On ne peut pas ne pas chercher le sens.

3. De ce qui fait sens.

Cette dynamique sémiotique fondamentale nous permettra d’introduire deux nouveaux facteurs, qui sont essentiels dans l’analyse des dimensions intentionnelles de la fonction sémiotique.

Il existe en effet une différence essentielle entre le sens (l’idée de la source d’une prégnance) et *ce qui fait venir vers nous* cette prégnance: la nature, une convention sociale ou culturelle, ou tel être particulier qui nous “fait signe”? Ce sont là des destinateurs que nous ne confondons pas dans la vie quotidienne, et qui font de nous des destinataires spécifiques: *observateurs* (ou admirateurs ou victimes) de la nature, *sujets* des instances socioculturelles, *partenaires* d’un individu expressif.

L’opération de contrôle qui produit la condition *p* (fig. 3) est vécue comme venant d’un destinateur, et nous amène ainsi à comprendre le sens dans son rapport à ce destinateur. Notre réponse, conduisant à la condition *q* (fig. 3), c’est-à-dire *notre intention d’aller vers le sens*, en tant que destinataires, se déroule dans la perspective d’une interaction volontaire et assumée avec ce destinateur.

Nous avons évoqué l’exemple de l’indice. Or, une analyse correspondante pourrait s’appliquer aux autres ordres sémiotiques, voire à l’ensemble des signes.

Qu’il soit trace, symptôme ou présage, l’indice peircien n’est pas très bavard: il ne dit pas en quoi un signe SA serait digne de l’intérêt soutenu du sujet. C’est que les *signes naturels* ne sont tels que grâce à quelque savoir prétabli dans l’esprit du sujet, portant sur le ‘flux’ causal ou méréologique du monde réel. Autrement dit, le sujet doit développer ou apprendre, puis connaître, ce lien RE-SA entre source et surgissement, et il doit posséder un programme d’action face à /SA/, dont l’objectivité ne révèle rien d’un tel programme subjectif. – En revanche, les *signes conventionnels* sont reconnus comme tels en fonction d’un code (avec inventaire et combinatoire) qui doit être acquis comme une compétence sociale par le sujet. Ils se distinguent des signes naturels par leurs liens RE-SA convenus et assumés par un contexte social. Ici, SA apparaît toujours à un endroit précis et spécifique dans l'espace social, dans un cadre de communication, et comme une saillance qui indique, de par sa propre configuration, qu'il est un *artéfact* posé là précisément par une autorité dans le but de signifier à un ensemble de sujets une instruction, au nom de quelque valeur conceptuelle (exemple: une stèle invitant à commémorer un passé historique spécifique). Le flux "afférent" de RE à SA est donc reçu par le sujet comme la voix du social (l'appel d'une autorité), et sa réponse interprétative, "efférente", est reçue par le contexte social, corrélativement, comme une obéïssance ou une désobéïssance au principe collectif qui avait fait entendre sa voix. Exemple: les drapeaux nationaux, les feux de circulation, les insignes, médailles, emblèmes, même les signes de l'écriture. – Finalement, les *signes intentionnels* au sens propre, ceux qui se négocient entre individus, et qui s'inventent au cours de la communication. Si les signes naturels correspondent a la catégorie consacrée des indices, et que les signes conventionnels correspondent à celle des symboles, cette dernière série s'avère reprendre et regrouper les propriétés de la catégorie des *icones*. Les images de toutes sortes, du portrait au diagramme, du plan géographique au modèle de l'architecte, et à l'oeuvre d'art, partagent cette condition iconique, dans la mesure où l'artéfact signifiant garde la plasticité de l'expression individuelle, et en même temps montre ce qu'il "veut dire" ans le code cognitif du sensible. Ce "code" du sensible diffère des codes sociaux par l'absence de rigidité; ainsi, dans le registre du langage, on peut dire que la *langue* est symbolique, alors que la *parole* est iconique. *Le code iconique du sensible*, de ce que nos sens arrivent à capter et retenir comme une forme, gestalt, chose figurative, est un principe fondé dans notre neuro-biologie, et responsable du fait que nous pouvons *représenter*. Car représenter est fondamentalement reconnaître une entité ou l'idée d'une entité par sa forme. Ainsi, une forme produite par un geste, ou une ligne graphique, une phrase musicale, peut être retenue et isolée dans le percept pour ensuite être reconnue, retrouvée dans d'autres percepts, de n'importe quelle nature. La *représentation* en général est possible, parce que le sensible est iconiquement codé, et que nos propres émissions expressives et intentionnelles s'en servent pour signifier. Si une saillance prégnante signifie par sa forme, reprenant la forme de l'entité visée, elle nous permet de communiquer sans faire le détour des conventions de la communauté; l'iconique peut donc se passer du symbolique, parce qu'il y a dans notre perception sensible une sensibilité aux formes qui active notre imaginaire. Si la communauté parle dans le symbolique, c'est au contraire l'individuel, la personne même qui fait signe, qui prend la parole dans l'iconique. Nous percevons assez bien cette distinction dans le quotidien – celui du travail, par excellence, où les deux registres arrivent souvent à se contredire.

Quel que soit le type de signes ou le registre sémiotique considéré, nous aboutissons au même résultat, à savoir que les deux aspects se rencontrent: le sens prégnant – envoyé par une instance, ou bien le monde physique (comme cause), le monde social (comme autorité), ou enfin l'autre individuel (comme être intentionnel) – afflue vers le sujet, qui à son tour répond par son attention, suivie d'une action (physique), d'une réponse (sociale), ou d'une réflexion (faite à l'autre).

Nous voudrions à cette occasion présenter un panorama classificatoire des signes – naturels, conventionnels ou intentionnels – faisant suite au tableau ternaire (icones, indices, symboles) qui est souvent donné comme fondamental, et faisant suite également aux considérations thomiennes sur le temps dans le renvoi sémiotique[6]. Nous distinguerons dans la dimension RE → SA les *renvois temporels*, c'est-à-dire ceux qui font signe vers le passé, le présent ou le futur; et dans la dimension /SA/ → RE^{p}, les *types de renvois* dont il peut s'agir, soit l'un des trois types de forces de signification connus jusqu'à notre présent: la *causalité* physique, la *convention déontique* socio-culturelle, arbitrairement donnée (typiquement afin de signifier un appel, une instruction), ou finalement la référence sémantique libre, *intentionnellement* établie lors d'une communication entre sujets co-présents et engagés dans une interaction réalisée face-à-face (qui fait partager un savoir, une hypothèse, une crainte, un espoir, un projet, etc.). On combine les renvois temporels avec les types de force de renvoi, et l'on obtient un tableau que nous croyons élémentaire. Les catégories générées par cette combinatoire simple correspondent en effet assez directement à certaines intuitions immédiates et à des distinctions reflétées par le langage (Fig. 4). Les termes sont bien entendu provisoires et ne servent qu'à donner une idée de la variante visée:

6 Cf. Thom (1980) encore.

force du signe	temps du contenu par rapport au présent de la prégnance:		
	Passé	Présent	Futur
NAT	*traces*	*symptômes*	*présage* – cf. indices
CON	*emblèmes*	*signaux*	*signes digitaux* – cf. symboles
INT	*images*	*plans*	*diagramme* – cf. icones

Légende: NAT = signes naturels, CON = signes conventionnels, INT = signes intentionnels.

Figure 3. Tableau élémentaire.

Un individu vit dans trois mondes: un monde naturel, un monde socio-culturel, et un monde intentionnel, plus intimiste, fait d'empathie et donc de contacts inter-humains directs; il vit aussi dans le présent, le passé et le futur. Les neuf catégories[7] qui s'ensuivent correspondent grosso modo aux modes sémiotiques connus.

Les symptômes renvoient à des états présents; les traces, par contre, aux états passés; et les présages aux états futurs.

Les signaux ordonnent des actes dans le présent (en indiquant ce qu'il faut faire, si l'on se trouve en présence du signe); les emblèmes renvoient au passé (à ce qu'il faut saluer et vénérer ou abhorrer dans la mémoire collective); et les symboles digitaux, normalement élaborés et utilisés à l'échelle de nos mains, à ce qu'un calcul numérique, une partition musicale ou un texte peut dire sur le monde à venir.

Ce qui nous mène au langage humain comme phénomène sémiotique particulier à notre espèce. Le langage (manifesté par les langues et leur usage) est conventionnel dans sa rigidité symbolique de structure; la phonétique, la grammaire et le lexique d'une langue ne sont certainement pas négotiables au niveau de la communication directe. On peut aussi faire remarquer que les trois registres temporels du signe symbolique correspondent aux trois registres des actes langagiers: le directif (renvoyant au présent), le commissif (renvoyant au futur) et le constatif (renvoyant au passé). Néanmoins, la pa-

7 Ce tableau élargit la fenêtre standard sur les classes des signes; il s'agit bien toujours d'un inventaire optimiste, qui se voudrait en principe exhaustif.

role est intentionnelle par excellence comme “vouloir dire”, et donc iconique. Elle fonctionne à chaque fois dans un échange où un énoncé est dit et compris comme un plan offert, une carte géographique et thématique d’orientation: “Vous êtes ici, et je vous dis, vu ce qui se passe entre vous et moi, que ce qui se passe là-bas, ou même un peu plus loin, va vous contraindre à X ou vous permettre d’Y...”. L’énonciation est déictique comme une carte; c’est le *hic et nunc* du présent étendu, pour ainsi dire. L’usage descriptif et narratif du discours est de l’ordre du portrait et de l’image en général, renvoyant au passé; seul l’usage argumentatif pointe vers l’avenir (la pensée est futurique, voire eschatologique, par vocation). Symbolique et iconique, le langage comme phénomène sémiotique ne manque pas non plus de manifester des traits indiciels. Ainsi, pour la philologie, les propriétés singulières d’une langue sont des traces du passé linguistique; les particularités d’un usage sont symptomatiques des êtres parlants, et dans plusieurs sens, par exemple sociologiquement ou psychologiquement; et le discours glossolalique des visionnaires, des poètes ou des possédés est traditionnellement reçu comme prophétique (le message est un présage). Le langage, au sens large, remplit donc tout le tableau des types principaux de signes.

Il va de soi que beaucoup de signes existants, sans doute potentiellement tous, fonctionnent dans plusieurs registres à la fois; on peut néanmoins distinguer des fonctions primaires, secondaires et éventuellement tertiaires, etc., selon la dominance relative des fonctions. C’est ainsi que, dans le langage verbal, une phrase prononcée – “Il pleut” – peut avoir, outre sa fonction dénotative directe (intentionnelle), une connotation codée (conventionnelle: prière de ne pas sortir, ou d’ouvrir le parapluie, ou de fermer la fenêtre) et un sens symptomatique, indiciel (naturel, p. ex.: hydrophobie ou hydrophilie de celui qui parle).

Dans les trois registres des signes – naturel, conventionnel, intentionnel – on peut à la limite distinguer un énonciateur et un énonciataire (les deux instances de l’énonciation). Ce sont les forces de contrôle du flux prégnanciel: la Nature, la Culture, et l’Autre en personne. L’attention de l’énonciataire est donc interpellée dans trois registres; en bref, par le destin, la morale et l’empathie. Le statut de communication au sens propre est réservé au dernier registre, intentionnel et iconique.

4. Quand dire, c'est faire

Les actes langagiers, symbolico-iconiques, c'est-à-dire conventionnellement intentionnels, consistent à signifier et, par là, à modifier, par le geste langagier même, C, la topologie sémantique de l'espace interhumain R que partagent les interlocuteurs. C affecte R qui affecte en retour C.

Il est facile de voir qu'une permission, proférée par un énonciateur, enlève une barrière entre l'énonciataire et un état désiré par celui-ci. Exemple (Fig. 4):

« Entrez, s'il vous plaît »

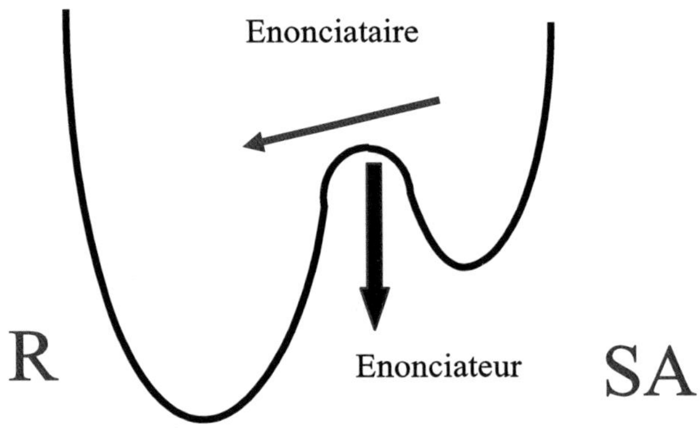

Figure 4. Dynamique du cusp, à deux minima. L'énonciateur baisse la barrière et invite son destinataire à aller de SA à RE, ce que ce dernier doit déjà être enclin à faire.

Tous les sens (directifs) du verbe *permettre* devrait pouvoir se modeler selon cette dynamique: baisser une barrière *pour que* le destinataire-énonciataire puisse passer, accéder à un état de référence valorisé par ce dernier.

En revanche, les actes langagiers (commissifs) impliquant un *obliger*, créant un *devoir*, éventuellement un "auto-devoir", comme la phrase suivante, ferment pour ainsi dire la porte *derrière* le sujet destinataire (Fig. 5):

« Je te promets de venir! »

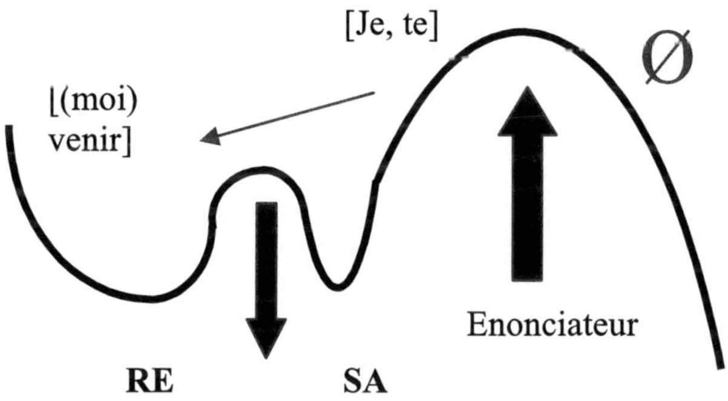

Figure 5. Dynamique queue d'aronde, à deux minima et un abîme.

On renforce la barrière entre SA et Ø, diminue celle qui sépare SA et RE, et l'on "force" ainsi un sujet de l'énonciation à faire qch. qu'il "doit" faire.

C'est peut-être cette capacité d'articuler ainsi l'intentionnel direct et dynamique, auto-référentiel: C → R → C, donc aussi l'émotionnel interpersonnel, non seulement le moral (conventionnel et normatif) ou l' "hystérique" (naturel, indiciel, affectif causé, corporel, "fou") qui confère au langage à la fois sa mauvaise réputation de flou conceptuel et sa bonne réputation de fondateur principal des cultures.[8]

5. Conclusion liquide

La confluence des deux bras du fleuve Thom, celui des catastrophes et celui des prégnances, en un seul cours nous donne un modèle sémiotique permettant de voir en quoi ce qui pourra s'appeler "référent", du côté de l'instance destinatrice, la *source de prégnance*, diffère de ce qu'on peut appeler le sens signifié, du côté d'un destinataire, à savoir la destination ou la vocation de la *prégnance même*, en tant que "signifiance"; mais elle signifie surtout que cette signifiance doit se donner dans la *scène* du signe, l'espace sémiotique de base où quelque chose de saillant deviendra signe (saillant prégnant) pour

8 Et c'est sans doute ce qui faisait dire à Roland Barthes, dans un discours célèbre, que le langage, comme tel, est autoritaire.

quelqu'un; que l'attention et l'intuition subjectives sont autant de "flux" phénoménologiques qui doivent couler vers le signe et au-delà du signe, vers le sens, idéalement dans le même lit que le flux qui avait créé du signifiant, et sans lequel il n'y aurait donc pas eu de signe.

Une telle sémiotique générale, faite de flux et de forces, plutôt que de catégories et de distinctions, nous semble particulièrement pertinente dans la perspective de la recherche cognitive des fondements de la conscience et de la pensée humaines, dans la mesure où elle rappelle notre base animale tout en invitant à bâtir sur cette base ce qu'il faut pour saisir sans réductionnisme les particularités de la signification culturelle.

References

Brandt, Per Aage (1993). *La charpente modale du sens. Pour une sémio-linguistique morphogénétique et dynamique*, John Benjamins and Aarhus University Press, Amsterdam & Aarhus.

Hjelmslev, Louis (1943/1993). *Omkring sprogteoriens grundlæggelse*. [Prolegomena to a Theory of Language]. Travaux du Cercle linguistique de Copenhague, Vol. XXV, The Linguistic Circle of Copenhagen, Copenhagen.

Thom, René (1980). "L'espace et les signes", in: *Semiotica* 29, 3&4: 193–208 (Conférence faite à Bloomington en 1979).

Thom, René (1991/2003). "Saillance et Prégnance", first published in: *L'inconscient et la Science*: 64–82, (ed. by R. Dorey), Dunod, Paris; citations are based on Thom, 2003.

Thom, René (1990). "Le problème des ontologies régionales en science", in: Apologie du logos, 1990, Hachette, Paris: 455–467.

Thom, René (2003). *Oeuvres complètes*, CD-Rom, Institut des Hautes Études Scientifiques, Paris Bures-sur-Yvette.

Contributors

Prof. Dr. PER AAGE BRANDT (*1944). Ph.D. in: Syntactic analysis and linguistic theory 1971; Habilitation (Dr. Phil., Docteur d'Etat, Sorbonne) in Semio-linguistics and the modal dynamics of meaning. 1975–2005 Professor at the University of Aarhus (AU); co-founder of the Center for Semiotics, at AU. 2005 – Professor of Cognitive Science and Modern Languages at Case Western Reserve University, Cleveland.

Major books: L'analyse phrastique (1973); La Charpente modale du sens (1992); Dynamiques du sens (1994); Morphologies of Meaning (1995); Spaces, Domains, and Meaning. Essays in Cognitive Semiotics (2004). Author of many articles and book chapters in: general and cultural semiotics, linguistic theory, cognitive semantics, poetics, musicology. Literary translator and writer; musician.

Associate Professor PEER F. BUNDGAARD (*1967). Ph.D. in General Semiotics 1996. Since 2005 Associate Professor at the Center for Semiotics, Aarhus University, Denmark.

Major publications: Kunst-Semiotiske beskrivelser af æstetisk oplevelse og betydning (2004), Kognitiv Semiotik (ed., 2003); articles and book chapters on cognitive semiotics, cognitive aesthetics, phenomenology and cognitive linguistics.

Prof. MARC CHAPERON (*1949) Doctorat d'Etat (habilitation) in Mathematics 1980. Since 1988 professor of Mathematics at the University of Paris 7.

Major book: Géométrie différentielle et singularités de systèmes dynamiques (1986). Author of many articles in Mathematics, a few in Philosophy or History of Science and one on Darius Milhaud and Robert Desnos.

Prof. ÁNGEL LÓPEZ–GARCÍA has taught at the universities of Zaragoza, Virginia, Mainz and Aarhus, and is actually professor of "Lingüística General" at the "Universidad de Valencia" (Spain). He received the price "Honoré Chaveé de l'Acadèmie des Belles Lettres" from the Institut de France for his scientific work.

Major publications: Gramática del español (1994, 1996, 1998), Fundamentos genéticos del lenguaje (2002), The grammar of genes (2005), Gramática cognitiva para profesores de español L2. (2005), The Neural Basis of Language (2007).

ISABEL MARCOS (*1964) is at present Senior Research fellow of e-GEO – Centre of Geography and Regional Development Studies at the New University of Lisbon. She is Doctor in Semiotics of the Architecture of the City at the University of Aarhus (Denmark) and Doctor in the Communication Sciences at the New University of Lisbon (Portugal).

Major books: As marcas da cultura mundializada na forma de Lisboa (2009). Dynamiques de la ville. Essais de sémiotique de l'espace (ed., 2007). Construir Cidade, Conquistar Espaço: Como é que uma sociedade pensa o seu futuro? (ed., 2007). Author of many articles, book chapters and papers to Conferences.

Associate Professor SVEND ØSTERGAARD (*1949). Ph.D. in General Semiotics 1994. Since 1998 Associate Professor at the Center for Semiotics, University of Aarhus, Denmark.

Major publications: The Mathematics of Meaning (1997), Kognition og Katastrofer. Studier i Dynamisk Semiotik (1998); articles on cognitive aesthetics, cognitive linguistics and catastrophe theory.

JEAN PETITOT (*1944) is Professor at the School for Advanced Studies in Social Sciences (EHESS) in Paris. He is a specialist of mathematical modeling in cognitive sciences and semiolinguistics (singularity theory and dynamical systems in René Thom's style).

He is the author of Morphogenèse du sens (Presses Universitaires de France, Paris, 1985; Peter Lang, Bern, 2004), Physique du Sens (Paris, CNRS, 1992), Neurogéométrie de la vision. Modèles mathématiques et physiques des architectures fonctionnelles (Editions de l'Ecole Polytechnique, Paris, 2008), and more than 300 papers.

FREDERIK STJERNFELT (*1957), *Professor at the* Center for Semiotics, University of Aarhus, graduated from the Nordic Department of University of Copenhagen with a degree in Nordic Literature and from 1992 to 2004 he was at the Dept. for Comparative Literature at the University of Copenhagen.

The inspiration from catastrophe theory and his general scientific outlook had already predisposed him to a realist view of semiotics, and Peirce's notion of diagrams caught his interest as a suitable meta-concept for such a realist semiotics. He decided to pick that issue for his dr. phil. treatise which was much later defended (2006) and published (2007) under the headline of "*Diagrammatology*".

Prof. Dr. WOLFGANG WILDGEN (*1944) Ph.D. in: General Linguistics and German Linguistics; Habilitation (Dr. phil. habil.) in General Linguistics 1981. Since 1981 professor of General and German Linguistics at the University of Bremen (emeritus since 2009).

Major books: Catastrophe Theoretic Semantics (1982); Process, Image, and Meaning (1994); De la grammaire au discours (1999); The Evolution of Human Language (2004); Kognitive Grammatik (2008); Die Sprachwissenschaft des 20. Jh.s (2010). Author of many articles and book chapters in: cognitive linguistics. philosophy of language, sociolinguistics, and semiotics.

European Semiotics: *Language, Cognition, and Culture*
Sémiotiques Européennes: *langage, cognition et culture*

Edited by / Série rédigée par
Per Aage Brandt (Cleveland), Wolfgang Wildgen (Bremen/Brême),
and/et Barend van Heusden (Groningen/Groningue)

European Semiotics originated from an initiative launched by a group of researchers in Semiotics from Denmark, Germany, Spain, France and Italy and was inspired by innovative impulses given by René Thom and his "semiophysics". The goal of the series is to provide a broad European forum for those interested in semiotic research focusing on *semiotic dynamics* and combining *cultural, linguistic and cognitive perspectives.*

This approach, which has its origins in Phenomenology, Gestalt Theory, Philosophy of Culture and Structuralism, views semiosis primarily as a cognitive process, which underlies and structures human culture. Semiotics is therefore considered to be the discipline suited *par excellence* to bridge the gap between the realms of the Cognitive Sciences and the Sciences of Culture.

The series publishes monographs, collected papers and conference proceedings of a high scholarly standard. Languages of publication are mainly English and French.

Sémiotiques européennes est le résultat d'une initiative prise par un groupe de chercheurs en sémiotique, originaires du Danemark, d'Allemagne, d'Espagne, de France et d'Italie, inspirée par l'impulsion innovatrice apportée par René Thom et sa "sémiophysique". Le but de cette collection est de fournir une tribune européenne large à tous ceux qui s'intéressent à la recherche sémiotique portant sur *les dynamiques sémiotiques,* et réunissant des *perspectives culturelles, linguistiques et cognitives.*

Cette approche, qui combine différentes sources, telle que la phénoménologie, le gestaltisme, la philosophie de la culture et le structuralisme, part du principe que la sémiosis est essentiellement un procès cognitif, qui sous-tend et structure toute culture humaine. Dans cette approche, la sémiotique est donc considérée comme la discipline par excellence capable de créer un pont entre les domaines des sciences cognitives et ceux des sciences de la culture.

Sémiotiques européennes accueille tant des monographies que des anthologies et des actes de colloques d'un haut niveau de recherche, rédigés de préférence en anglais et en français.

Volume 1	Wolfgang Wildgen: De la grammaire au discours Une approche morphodynamique xi + 351 pp. 1999. ISBN 3-906762-03-3 / US-ISBN 0-8204-4226-7
Volume 2	Lene Fogsgaard: Esquemas copulativos de SER y ESTAR Ensayo de semiolingüística 382 pp. 2000. ISBN 3-906764-22-2
Volume 3	Jean Petitot: Morphogenesis of Meaning xii + 279 pp. 2003. ISBN 3-03910-104-8 / US-ISBN 0-8204-6858-4
Volume 4	Per Aage Brandt: Spaces, Domains, and Meaning Essays in Cognitive Semiotics 271 pp. 2004. ISBN 3-03910-227-3 / US-ISBN 0-8204-7006-6
Volume 5	Marcel Bax, Barend van Heusden & Wolfgang Wildgen: Semiotic Evolution and the Dynamics of Culture xx + 318 pp. 2004. ISBN 3-03910-394-6 / US-ISBN 0-8204-9
Volume 6	Ángel López-García: The Grammar of Genes How the Genetic Code Resembles the Linguistic Code 182 pp. 2005. ISBN 3-03910-654-6 / US-ISBN 3-03910-7171-2
Volume 7	Ole Kühl: Musical Semantics 261 pp. 2nd unmodified printing. 2007, 2008. ISBN 978-3-03911-718-5
Volume 8	Todd Oakley: From Attention to Meaning Explorations in Semiotics, Linguistics, and Rhetoric 268 pp. 2009. ISBN 978-3-03911-442-9
Volume 9	Wolfgang Wildgen & Barend van Heusden (eds.): Metarepresentation, Self-Organization and Art 298 pp. 2009. ISBN 978-3-03911-684-3
Volume 10	Wolfgang Wildgen & Per Aage Brandt (eds.): Semiosis and Catastrophes René Thom's Semiotic Heritage 185 pp. 2010. ISBN 978-3-0343-0467-2